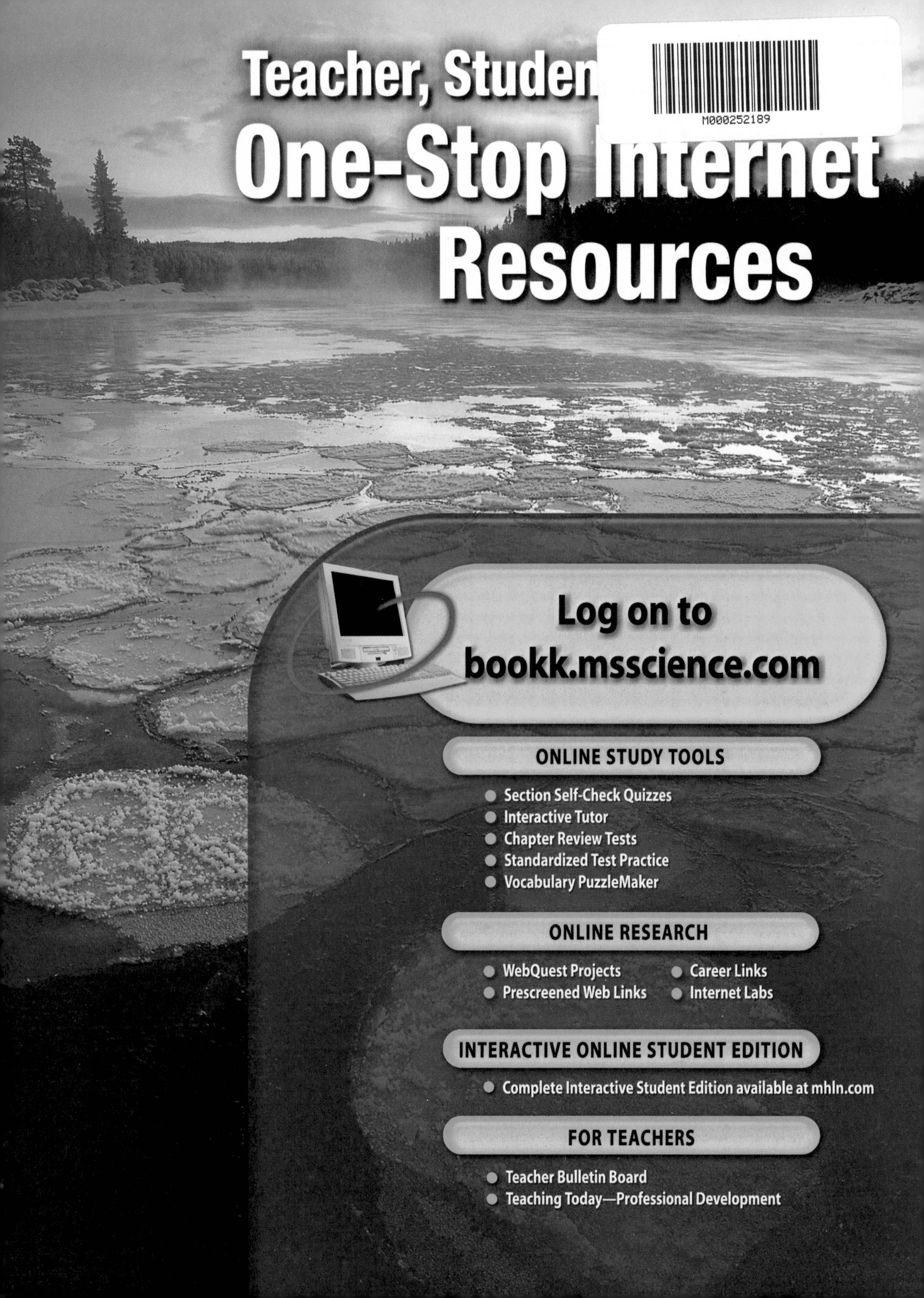

Teacher, Student
One-Stop Internet Resources

Log on to
bookk.msscience.com

ONLINE STUDY TOOLS

- Section Self-Check Quizzes
- Interactive Tutor
- Chapter Review Tests
- Standardized Test Practice
- Vocabulary PuzzleMaker

ONLINE RESEARCH

- WebQuest Projects
- Prescreened Web Links
- Career Links
- Internet Labs

INTERACTIVE ONLINE STUDENT EDITION

- Complete Interactive Student Edition available at mhln.com

FOR TEACHERS

- Teacher Bulletin Board
- Teaching Today—Professional Development

SAFETY SYMBOLS	HAZARD	EXAMPLES	PRECAUTION	REMEDY
DISPOSAL	Special disposal procedures need to be followed.	certain chemicals, living organisms	Do not dispose of these materials in the sink or trash can.	Dispose of wastes as directed by your teacher.
BIOLOGICAL	Organisms or other biological materials that might be harmful to humans	bacteria, fungi, blood, unpreserved tissues, plant materials	Avoid skin contact with these materials. Wear mask or gloves.	Notify your teacher if you suspect contact with material. Wash hands thoroughly.
EXTREME TEMPERATURE	Objects that can burn skin by being too cold or too hot	boiling liquids, hot plates, dry ice, liquid nitrogen	Use proper protection when handling.	Go to your teacher for first aid.
SHARP OBJECT	Use of tools or glassware that can easily puncture or slice skin	razor blades, pins, scalpels, pointed tools, dissecting probes, broken glass	Practice common-sense behavior and follow guidelines for use of the tool.	Go to your teacher for first aid.
FUME	Possible danger to respiratory tract from fumes	ammonia, acetone, nail polish remover, heated sulfur, moth balls	Make sure there is good ventilation. Never smell fumes directly. Wear a mask.	Leave foul area and notify your teacher immediately.
ELECTRICAL	Possible danger from electrical shock or burn	improper grounding, liquid spills, short circuits, exposed wires	Double-check setup with teacher. Check condition of wires and apparatus.	Do not attempt to fix electrical problems. Notify your teacher immediately.
IRRITANT	Substances that can irritate the skin or mucous membranes of the respiratory tract	pollen, moth balls, steel wool, fiberglass, potassium permanganate	Wear dust mask and gloves. Practice extra care when handling these materials.	Go to your teacher for first aid.
CHEMICAL	Chemicals can react with and destroy tissue and other materials	bleaches such as hydrogen peroxide; acids such as sulfuric acid, hydrochloric acid; bases such as ammonia, sodium hydroxide	Wear goggles, gloves, and an apron.	Immediately flush the affected area with water and notify your teacher.
TOXIC	Substance may be poisonous if touched, inhaled, or swallowed.	mercury, many metal compounds, iodine, poinsettia plant parts	Follow your teacher's instructions.	Always wash hands thoroughly after use. Go to your teacher for first aid.
FLAMMABLE	Flammable chemicals may be ignited by open flame, spark, or exposed heat.	alcohol, kerosene, potassium permanganate	Avoid open flames and heat when using flammable chemicals.	Notify your teacher immediately. Use fire safety equipment if applicable.
OPEN FLAME	Open flame in use, may cause fire.	hair, clothing, paper, synthetic materials	Tie back hair and loose clothing. Follow teacher's instruction on lighting and extinguishing flames.	Notify your teacher immediately. Use fire safety equipment if applicable.

 Eye Safety Proper eye protection should be worn at all times by anyone performing or observing science activities.

 Clothing Protection This symbol appears when substances could stain or burn clothing.

 Animal Safety This symbol appears when safety of animals and students must be ensured.

 Handwashing After the lab, wash hands with soap and water before removing goggles.

Glencoe Science

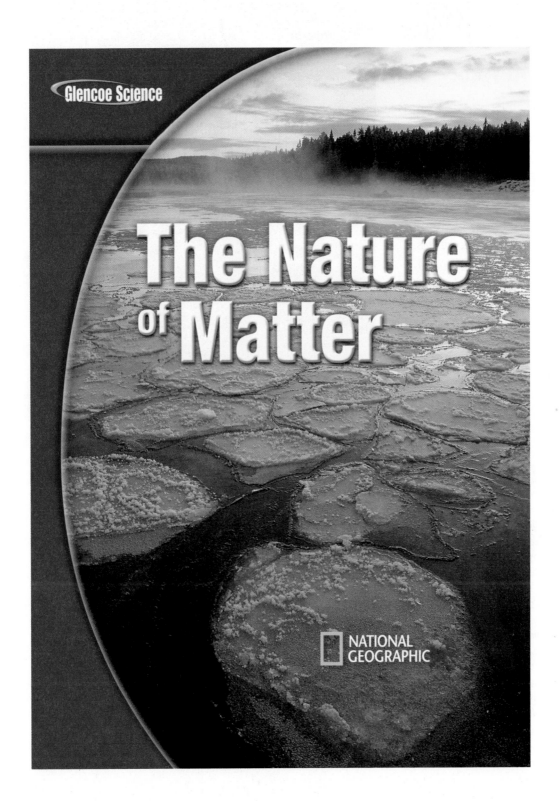

The Nature of Matter

NATIONAL GEOGRAPHIC

Mc Graw Hill **Glencoe**

New York, New York   Columbus, Ohio   Chicago, Illinois   Woodland Hills, California

Glencoe Science

The Nature of Matter

This pancake ice has formed on a river in Sweden. Pancake ice forms when surface slush, arising from snow falling on water that is already at the freezing temperature, freezes. The surface slush collects into rounded floating pads that collide and separate.

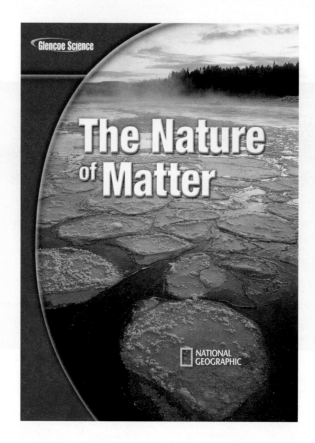

 **Glencoe**

The *McGraw-Hill* Companies

Send all inquiries to:
Glencoe/McGraw-Hill
8787 Orion Place
Columbus, OH 43240-4027

ISBN: 978-0-07-877832-2
MHID: 0-07-877832-8

Printed in the United States of America.

4 5 6 7 8 9 10 DOW 12 11 10 09

Authors

NATIONAL GEOGRAPHIC
Education Division
Washington, D.C.

Eric Werwa, PhD
Department of Physics and Astronomy
Otterbein College
Westerville, OH

Patricia Horton
Mathematics and Science Teacher
Summit Intermediate School
Etiwanda, CA

Dinah Zike
Educational Consultant
Dinah-Might Activities, Inc.
San Antonio, TX

Thomas McCarthy, PhD
Science Department Chair
St. Edward's School
Vero Beach, FL

Series Consultants

CONSULTANTS

Jack Cooper
Ennis High School
Ennis, TX

Linda McGaw
Science Program Coordinator
Advanced Placement Strategies, Inc.
Dallas, TX

MATH

Michael Hopper, DEng
Manager of Aircraft Certification
L-3 Communications
Greenville, TX

READING

Barry Barto
Special Education Teacher
John F. Kennedy Elementary
Manistee, MI

SAFETY

Aileen Duc, PhD
Science 8 Teacher
Hendrick Middle School, Plano ISD
Plano, TX

Sandra West, PhD
Department of Biology
Texas State University-San Marcos
San Marcos, TX

ACTIVITY TESTERS

Nerma Coats Henderson
Pickerington Lakeview Jr. High
School
Pickerington, OH

Mary Helen Mariscal-Cholka
William D. Slider Middle School
El Paso, TX

Science Kit and Boreal Laboratories
Tonawanda, NY

Series Reviewers

Sharla Adams
IPC Teacher
Allen High School
Allen, TX

Anthony J. DiSipio, Jr.
8th Grade Science
Octorana Middle School
Atglen, PA

Sandra Everhart
Dauphin/Enterprise Jr. High Schools
Enterprise, AL

George Gabb
Great Bridge Middle School
Chesapeake Public Schools
Chesapeake, VA

Michelle Mazeika-Simmons
Whiting Middle School
Whiting, IN

HOW TO...

Use Your Science Book

Why do I need my science book?

Have you ever been in class and not understood all of what was presented? Or, you understood everything in class, but at home, got stuck on how to answer a question? Maybe you just wondered when you were ever going to use this stuff?

These next few pages are designed to help you understand everything your science book can be used for . . . besides a paperweight!

Before You Read

- **Chapter Opener** Science is occurring all around you, and the opening photo of each chapter will preview the science you will be learning about. The **Chapter Preview** will give you an idea of what you will be learning about, and you can try the **Launch Lab** to help get your brain headed in the right direction. The **Foldables** exercise is a fun way to keep you organized.

- **Section Opener** Chapters are divided into two to four sections. The **As You Read** in the margin of the first page of each section will let you know what is most important in the section. It is divided into four parts. **What You'll Learn** will tell you the major topics you will be covering. **Why It's Important** will remind you why you are studying this in the first place! The **Review Vocabulary** word is a word you already know, either from your science studies or your prior knowledge. The **New Vocabulary** words are words that you need to learn to understand this section. These words will be in **boldfaced** print and highlighted in the section. Make a note to yourself to recognize these words as you are reading the section.

Glencoe Science

The Nature of Matter

NATIONAL GEOGRAPHIC

As You Read

- **Headings** Each section has a title in large red letters, and is further divided into blue titles and small red titles at the beginnings of some paragraphs. To help you study, make an outline of the headings and subheadings.

- **Margins** In the margins of your text, you will find many helpful resources. The **Science Online** exercises and **Integrate** activities help you explore the topics you are studying. **MiniLabs** reinforce the science concepts you have learned.

- **Building Skills** You also will find an **Applying Math** or **Applying Science** activity in each chapter. This gives you extra practice using your new knowledge, and helps prepare you for standardized tests.

- **Student Resources** At the end of the book you will find **Student Resources** to help you throughout your studies. These include **Science, Technology,** and **Math Skill Handbooks,** an **English/Spanish Glossary,** and an **Index.** Also, use your **Foldables** as a resource. It will help you organize information, and review before a test.

- **In Class** Remember, you can always ask your teacher to explain anything you don't understand.

FOLDABLES™
Study Organizer

Science Vocabulary Make the following Foldable to help you understand the vocabulary terms in this chapter.

STEP 1 Fold a vertical sheet of notebook paper from side to side.

STEP 2 Cut along every third line of only the top layer to form tabs.

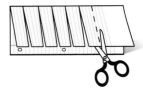

STEP 3 Label each tab with a vocabulary word from the chapter.

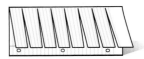

Build Vocabulary As you read the chapter, list the vocabulary words on the tabs. As you learn the definitions, write them under the tab for each vocabulary word.

Look For...
FOLDABLES™
At the beginning of every section.

In Lab

Working in the laboratory is one of the best ways to understand the concepts you are studying. Your book will be your guide through your laboratory experiences, and help you begin to think like a scientist. In it, you not only will find the steps necessary to follow the investigations, but you also will find helpful tips to make the most of your time.

● Each lab provides you with a **Real-World Question** to remind you that science is something you use every day, not just in class. This may lead to many more questions about how things happen in your world.

● Remember, experiments do not always produce the result you expect. Scientists have made many discoveries based on investigations with unexpected results. You can try the experiment again to make sure your results were accurate, or perhaps form a new hypothesis to test.

● Keeping a **Science Journal** is how scientists keep accurate records of observations and data. In your journal, you also can write any questions that may arise during your investigation. This is a great method of reminding yourself to find the answers later.

Look For...
● **Launch Labs** start every chapter.
● **MiniLabs** in the margin of each chapter.
● **Two Full-Period Labs** in every chapter.
● **EXTRA Try at Home Labs** at the end of your book.
● the **Web site** with **laboratory demonstrations**.

Before a Test

Admit it! You don't like to take tests! However, there *are* ways to review that make them less painful. Your book will help you be more successful taking tests if you use the resources provided to you.

- Review all of the **New Vocabulary** words and be sure you understand their definitions.

- Review the notes you've taken on your **Foldables,** in class, and in lab. Write down any question that you still need answered.

- Review the **Summaries** and **Self Check questions** at the end of each section.

- Study the concepts presented in the chapter by reading the **Study Guide** and answering the questions in the **Chapter Review.**

Look For...

- **Reading Checks** and **caption questions** throughout the text.
- the **Summaries** and **Self Check questions** at the end of each section.
- the **Study Guide** and **Review** at the end of each chapter.
- the **Standardized Test Practice** after each chapter.

Let's Get Started

To help you find the information you need quickly, use the Scavenger Hunt below to learn where things are located in Chapter 1.

1. What is the title of this chapter?

2. What will you learn in Section 1?

3. Sometimes you may ask, "Why am I learning this?" State a reason why the concepts from Section 2 are important.

4. What is the main topic presented in Section 2?

5. How many reading checks are in Section 1?

6. What is the Web address where you can find extra information?

7. What is the main heading above the sixth paragraph in Section 2?

8. There is an integration with another subject mentioned in one of the margins of the chapter. What subject is it?

9. List the new vocabulary words presented in Section 2.

10. List the safety symbols presented in the first Lab.

11. Where would you find a Self Check to be sure you understand the section?

12. Suppose you're doing the Self Check and you have a question about concept mapping. Where could you find help?

13. On what pages are the Chapter Study Guide and Chapter Review?

14. Look in the Table of Contents to find out on which page Section 2 of the chapter begins.

15. You complete the Chapter Review to study for your chapter test. Where could you find another quiz for more practice?

Teacher Advisory Board

The Teacher Advisory Board gave the editorial staff and design team feedback on the content and design of the Student Edition. They provided valuable input in the development of the 2008 edition of *Glencoe Science.*

John Gonzales
Challenger Middle School
Tucson, AZ

Rachel Shively
Aptakisic Jr. High School
Buffalo Grove, IL

Roger Pratt
Manistique High School
Manistique, MI

Kirtina Hile
Northmor Jr. High/High School
Galion, OH

Marie Renner
Diley Middle School
Pickerington, OH

Nelson Farrier
Hamlin Middle School
Springfield, OR

Jeff Remington
Palmyra Middle School
Palmyra, PA

Erin Peters
Williamsburg Middle School
Arlington, VA

Rubidel Peoples
Meacham Middle School
Fort Worth, TX

Kristi Ramsey
Navasota Jr. High School
Navasota, TX

Student Advisory Board

The Student Advisory Board gave the editorial staff and design team feedback on the design of the Student Edition. We thank these students for their hard work and creative suggestions in making the 2008 edition of *Glencoe Science* student friendly.

Jack Andrews
Reynoldsburg Jr. High School
Reynoldsburg, OH

Peter Arnold
Hastings Middle School
Upper Arlington, OH

Emily Barbe
Perry Middle School
Worthington, OH

Kirsty Bateman
Hilliard Heritage Middle School
Hilliard, OH

Andre Brown
Spanish Emersion Academy
Columbus, OH

Chris Dundon
Heritage Middle School
Westerville, OH

Ryan Manafee
Monroe Middle School
Columbus, OH

Addison Owen
Davis Middle School
Dublin, OH

Teriana Patrick
Eastmoor Middle School
Columbus, OH

Ashley Ruz
Karrer Middle School
Dublin, OH

The Glencoe middle school science Student Advisory Board taking a timeout at COSI, a science museum in Columbus, Ohio.

Contents

**Nature of Science:
Pencils into Diamonds—2**

chapter 1

Atoms, Elements, Compounds, and Mixtures—6

Section 1 **Models of the Atom** 8
Section 2 **The Simplest Matter** 18
    Lab Elements and the Periodic Table 24
Section 3 **Compounds and Mixtures** 25
    Lab: Design Your Own
        Mystery Mixture 30

chapter 2

States of Matter—38

Section 1 **Matter** 40
Section 2 **Changes of State** 45
    Lab The Water Cycle 53
Section 3 **Behavior of Fluids** 54
    Lab: Design Your Own
        Design Your Own Ship 62

In each chapter, look for these opportunities for review and assessment:
- Reading Checks
- Caption Questions
- Section Review
- Chapter Study Guide
- Chapter Review
- Standardized Test Practice
- Online practice at bookk.msscience.com

Get Ready to Read Strategies
- Preview 8A
- Monitor 40A
- Identify the Main Idea 72A
- Make Connections 98A

chapter 3

Properties and Changes of Matter—70

Section 1   **Physical and Chemical Properties** 72

    Lab   Finding the Difference 77

Section 2   **Physical and Chemical Changes** 78

    Lab: Design Your Own

      Battle of the Toothpastes 88

chapter 4

The Periodic Table—96

Section 1   **Introduction to the Periodic Table** 98

Section 2   **Representative Elements** 105

Section 3   **Transition Elements** 112

    Lab   Metals and Nonmetals 117

    Lab: Use the Internet

      Health Risks from Heavy Metals 118

Student Resources

Science Skill Handbook—128

  Scientific Methods 128

  Safety Symbols 137

  Safety in the Science

    Laboratory 138

Extra Try at Home Labs—140

Technology Skill Handbook—142

  Computer Skills 142

  Presentation Skills 145

Math Skill Handbook—146

  Math Review 146

  Science Applications 156

Reference Handbooks—161

  Physical Science

    Reference Tables 161

  Periodic Table of

    the Elements 162

English/Spanish Glossary—165

Index—170

Credits—174

Cross-Curricular Readings/Labs

DVD available as a video lab

NATIONAL GEOGRAPHIC VISUALIZING

1 The Periodic Table 20
2 States of Matter 48
3 Recycling 86
4 Synthetic Elements 115

TIME SCIENCE AND Society

1 Ancient Views of Matter 32

Oops! Accidents in SCIENCE

2 The Incredible Stretching Goo . . . 64

Science and Language Arts

15 "Anansi Tries to Steal All the
Wisdom in the World" 120

SCIENCE Stats

3 Strange Changes 90

Launch LAB

1 Model the Unseen 7
2 Experiment with a
Freezing Liquid 39
3 The Changing Face
of a Volcano 71
DVD **4** Make a Model of a
Periodic Pattern 97

Mini LAB

1 Comparing Compounds 26
2 Observing Vaporization 50
3 Measuring Properties 74
3 Identifying an
Unknown Substance 75
4 Designing a Periodic Table 99

Mini LAB Try at Home

1 Modeling the Nuclear Atom 15
2 Predicting a Waterfall 57
3 Comparing Changes 81

One-Page Labs

1 Elements and the
Periodic Table 24
2 The Water Cycle 53
3 Finding the Difference 77
4 Metals and Nonmetals 117

Design Your Own Labs

DVD **1** Mystery Mixture 30–31
DVD **2** Design Your Own Ship 62–63
DVD **3** Battle of the Toothpastes 88–89

Use the Internet Labs

4 Health Risks from
Heavy Metals 118–119

Applying Math

2 Calculating Density 59
3 Converting Temperatures 84

Applying Science

1 What's the best way to desalt
ocean water? 27
2 How can ice save oranges? 49
4 What does *periodic* mean
in the periodic table? 103

INTEGRATE

Astronomy: 83
Career: 108
Chemistry: 120
Earth Science: 29
Health: 116
History: 19, 42
Life Science: 28, 61, 81, 109
Physics: 16, 46, 114

Science Online

28, 43, 49, 51, 61, 76, 81, 102, 116

Standardized Test Practice

36–37, 68–69, 94–95, 124–125

Pencils into Diamonds

Diamond, the hardest mineral, is both beautiful and strong. Diamonds can cut steel, conduct heat, and withstand boiling acid. Unfortunately, to find a one carat gem-quality diamond, an average of 250 tons of rock must be mined!

But what if there were another way to get gems? In 1902, Auguste Verneuil, a French scientist, created the world's first synthetic ruby by carefully heating aluminum oxide powder. When other elements were added to this mixture, other colored gemstones were created.

Natural Diamonds

During World War II (1939–1945), there was a sudden need for hard gems used in the manufacturing of precision instruments. Around this time, scientists made the first diamond from carbon, or graphite—the same substance that is in #2 pencils. Graphite is made up of sheets of well-bonded carbon atoms. However, the sheets are only loosely bonded together. This gives graphite its flaky, slippery quality. Diamond, however, is made up of carbon atoms bonded strongly in three dimensions.

Figure 1 Uncut diamonds are ground and shaped into highly prized gems.

Figure 2 **A** Graphite has a layered structure of carbon atoms. Strong bonds exist within the layers and weak bonds exist between the layers. **B** All bonds between carbon atoms in diamond are strong.

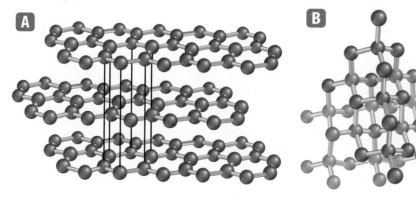

Graphite

Diamond

Making Synthetic Diamond

To change graphite to diamond, scientists expose it to extreme pressures and to temperatures as high as 3,000°C. The first experiments were unsuccessful. Scientists then reasoned that since diamond is a crystal, it might grow out of a super-concentrated solution as other crystals do. To dissolve carbon, they added melted troilite to their experiments. Troilite is a metal found surrounding tiny diamonds at meteorite impact sites. Finally, they succeeded. The first synthetic diamonds were yellowish, but they could be used in industry.

Diamond is a valuable material for industry. It is used to make machine-tool coatings, contact lenses, and electrodes. Computer engineers expect that diamond will soon be used to make high-speed computer chips.

Telling Them Apart

Of course, no one has forgotten the diamond's first use: decoration. The first synthetic diamonds were flawed by tiny pieces of metal from the diamond-making process and yellowed by the nitrogen in our atmosphere. Today, diamond makers have eliminated many of these problems. In just five days, labs now produce colorless, jewelry-grade synthetic diamonds that are much less expensive than natural diamonds. One natural diamond company now determines which stones are synthetic by using phosphorescence. Unlike natural diamonds, synthetic diamonds will glow in the dark for a few seconds after being exposed to ultraviolet light. Synthetic diamond makers are already working to eliminate this difference, too.

Some people are excited about affordable diamonds. Others are concerned that synthetic gem quality diamonds will be sold as natural diamonds. Some people also wonder if a diamond that was made in a laboratory in a few days has the same symbolic significance as a diamond formed naturally over millions of years.

Figure 4 A machine like this one can produce the temperature and pressure required to create a diamond.

Figure 3 Diamond is sought after for its beauty and for its useful properties.

Figure 5 The need for diamond-coated drill bits and cutting blades sparked research into the creation of gem-quality diamonds.

Science

The path to tomorrow's high-speed diamond computer chips will have begun with people trying to make jewelry! Such pathways are reminders of how science touches many aspects of human life. Like the pieces in a puzzle, each scientific breakthrough reveals more about how the world works.

Physical science includes the chemistry that produces synthetic diamonds. In this book, you'll learn how the elements on the periodic table combine to make up everything you see around you. You'll also learn how chemistry can change many aspects of the world around you.

Science Today

Scientists work to find solutions and to answer questions. As people's needs change, scientists who are developing new technologies change the direction of their work. For example, the need for diamonds during World War II triggered research into making synthetic diamonds.

Where Do Today's Scientists Work?

Scientists today work in a variety of places for a variety of reasons. Both scientists who study natural diamonds and people who make synthetic diamonds might work in controlled laboratory environments. They may also study diamonds where they are found in nature.

Public and Private Research

The United States government supports a great deal of scientific research. Publicly funded research usually deals with topics that affect the health and welfare of the country's citizens.

In the private sector, many companies, large and small, have their own laboratories. Their scientists research new technologies, use the technologies in the products that they sell, and test the new products. The world's first synthetic diamond was created by a private company that needed diamond for its products. Another private company, a diamond company, has created many of the world's synthetic diamonds in its laboratory! Why? The diamond company wants to understand how synthetic diamonds are made so that they can see the differences between their naturally formed diamonds and their competitors' manufactured diamonds. Research has helped them to develop a machine that identifies synthetic diamonds.

Research at Universities

Major universities also have laboratories. Their work with the government or corporations allows academic and industrial scientists to learn from one another. Industry and the government also provide grants and funding for university laboratories.

Dr. Rajiv K. Singh is a professor at the University of Florida, Gainesville. He and his colleague James Adair created the world's largest synthetic diamond using a process called chemical vapor deposition (CVD). Dr. Singh researches many different materials for the University. His work with synthetic diamonds also involves research in flat-panel displays, thin film batteries, electronics, and superconductors.

Figure 6 Researchers at the University of Florida created the world's largest synthetic diamond.

You Do It

You probably have many devices at home that new discoveries in science have made possible. For example, DVD and MP3 players are technologies that didn't exist just a few years ago. Research the science behind your favorite "gadget" and explain to the class how it works.

The BIG Idea

Atoms are modeled and classified to help people study and understand them.

SECTION 1
Models of the Atom
Main Idea Scientists use models to help understand the structure of atoms.

SECTION 2
The Simplest Matter
Main Idea Atoms have unique characteristics that are used to classify them.

SECTION 3
Compounds and Mixtures
Main Idea Atoms can form compounds and mixtures.

Atoms, Elements, Compounds, and Mixtures

What an impressive sight!

Have you ever seen iron on an atomic level? This is an image of 48 iron atoms surrounding a single copper atom. In this chapter, you will learn about scientists and their discoveries about the nature of the atom.

Science Journal Based on your knowledge, describe what an atom is.

Start-Up Activities

Model the Unseen

Have you ever had a wrapped birthday present that you couldn't wait to open? What did you do to try to figure out what was in it? The atom is like that wrapped present. You want to investigate it, but you cannot see it easily.

1. Your teacher will give you a piece of clay and some pieces of metal. Count the pieces of metal.
2. Bury these pieces in the modeling clay so they can't be seen.
3. Exchange clay balls with another group.
4. With a toothpick, probe the clay to find out how many pieces of metal are in the ball and what shape they are.
5. **Think Critically** In your Science Journal, sketch the shapes of the metal pieces as you identify them. How does the number of pieces you found compare with the number that were in the clay ball? How do their shapes compare?

Parts of the Atom Make the following Foldable to help you organize your thoughts and review parts of an atom.

STEP 1 **Collect** two sheets of paper and layer them about 1.25 cm apart vertically. Keep the edges level.

STEP 2 **Fold** up the bottom edges of the paper to form four equal tabs.

STEP 3 **Fold** the papers and crease well to hold the tabs in place. Staple along the fold. Label the flaps *Atom, Electron, Proton,* and *Neutron* as shown.

Atom
Electron
Proton
Neutron

Read and Write As you read the chapter, describe how each part of the atom was discovered and record other facts under the flaps.

Preview this chapter's content and activities at
bookk.msscience.com

Get Ready to Read

① Learn It! If you know what to expect before reading, it will be easier to understand ideas and relationships presented in the text. Follow these steps to preview your reading assignments.

1. Look at the title and any illustrations that are included.
2. Read the headings, subheadings, and anything in bold letters.
3. Skim over the passage to see how it is organized. Is it divided into many parts?
4. Look at the graphics—pictures, maps, or diagrams. Read their titles, labels, and captions.
5. Set a purpose for your reading. Are you reading to learn something new? Are you reading to find specific information?

② Practice It! Take some time to preview this chapter. Skim all the main headings and subheadings. With a partner, discuss your answers to these questions.
- Which part of this chapter looks most interesting to you?
- Are there any words in the headings that are unfamiliar to you?
- Choose one of the section review questions to discuss with a partner.

③ Apply It! Now that you have skimmed the chapter, write a short paragraph describing one thing you want to learn from this chapter.

Target Your Reading

Reading Tip

Preview a section or chapter before you read it by looking at the figures and captions. Then, you will have an idea what the text is about before you read it.

Use this to focus on the main ideas as you read the chapter.

1. **Before you read** the chapter, respond to the statements below on your worksheet or on a numbered sheet of paper.
   - Write an **A** if you **agree** with the statement.
   - Write a **D** if you **disagree** with the statement.

2. **After you read** the chapter, look back to this page to see if you've changed your mind about any of the statements.
   - If any of your answers changed, explain why.
   - Change any false statements into true statements.
   - Use your revised statements as a study guide.

Science Online
Print out a worksheet of this page at bookk.msscience.com

Before You Read A or D		Statement	After You Read A or D
	1	An element is composed of more than one type of atom.	
	2	According to John Dalton, an atom was a tiny sphere that was the same throughout.	
	3	Atoms contain protons, neutrons, and electrons.	
	4	The periodic table organizes compounds by their properties.	
	5	The atomic number is the number of protons in the nucleus of an atom.	
	6	Elements are classified as metals, nonmetals, or metalloids.	
	7	Nonmetals are good conductors of heat and electricity.	
	8	Compounds are composed of more than one element.	
	9	Air is an example of a compound.	

Models of the Atom

as you read

What You'll Learn

- **Explain** how scientists discovered subatomic particles.
- **Explain** how today's model of the atom developed.
- **Describe** the structure of the nuclear atom.

Why It's Important

All matter is made up of atoms. Atoms make up everything in your world.

⦿ Review Vocabulary

matter: anything that has mass and takes up space

New Vocabulary

- ● element
- ● electron
- ● proton
- ● neutron
- ● electron cloud

First Thoughts

Do you like mysteries? Are you curious? Humans are curious. Someone always wants to know something that is not easy to detect or to see what can't be seen. For example, people began wondering about matter more than 2,500 years ago. Some of the early philosophers thought that matter was composed of tiny particles. They reasoned that you could take a piece of matter, cut it in half, cut the half piece in half again, and continue to cut again and again. Eventually, you wouldn't be able to cut any more. You would have only one particle left. They named these particles *atoms*, a term that means "cannot be divided." Another way to imagine this is to picture a string of beads like the one shown in **Figure 1.** If you keep dividing the string into pieces, you eventually come to one single bead.

Describing the Unseen The early philosophers didn't try to prove their theories by doing experiments as scientists now do. Their theories were the result of reasoning, debating, and discussion—not of evidence or proof. Today, scientists will not accept a theory that is not supported by experimental evidence. But even if these philosophers had experimented, they could not have proven the existence of atoms. People had not yet discovered much about what is now called chemistry, the study of matter. The kind of equipment needed to study matter was a long way from being invented. Even as recently as 500 years ago, atoms were still a mystery.

Figure 1 You can divide this string of beads in half, and in half again until you have one, indivisible bead. Like this string of beads, all matter can be divided until you reach one basic particle, the atom.

A Model of the Atom

A long period passed before the theories about the atom were developed further. Finally during the eighteenth century, scientists in laboratories, like the one on the left in **Figure 2,** began debating the existence of atoms once more. Chemists were learning about matter and how it changes. They were putting substances together to form new substances and taking substances apart to find out what they were made of. They found that certain substances couldn't be broken down into simpler substances. Scientists came to realize that all matter is made up of elements. An **element** is matter made of atoms of only one kind. For example, iron is an element made of iron atoms. Silver, another element, is made of silver atoms. Carbon, gold, and oxygen are other examples of elements.

Dalton's Concept John Dalton, an English schoolteacher in the early nineteenth century, combined the idea of elements with the earlier theory of the atom. He proposed the following ideas about matter: (1) Matter is made up of atoms, (2) atoms cannot be divided into smaller pieces, (3) all the atoms of an element are exactly alike, and (4) different elements are made of different kinds of atoms. Dalton pictured an atom as a hard sphere that was the same throughout, something like a tiny marble. A model like this is shown in **Figure 3.**

Scientific Evidence Dalton's theory of the atom was tested in the second half of the nineteenth century. In 1870, the English scientist William Crookes did experiments with a glass tube that had almost all the air removed from it. The glass tube had two pieces of metal called electrodes sealed inside. The electrodes were connected to a battery by wires.

Figure 2 Even though the laboratories of the time were simple compared to those of today, incredible discoveries were made during the eighteenth century.

Figure 3 Dalton pictured the atom as a hard sphere that was the same throughout.
Describe *Dalton's theory of the atom.*

Figure 4 Crookes used a glass tube containing only a small amount of gas. When the glass tube was connected to a battery, something flowed from the negative electrode (cathode) to the positive electrode (anode).
Explain *if this unknown thing was light or a stream of particles.*

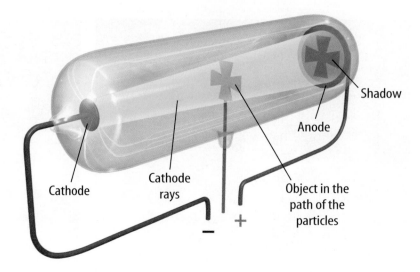

A Strange Shadow An electrode is a piece of metal that can conduct electricity. One electrode, called the anode, has a positive charge. The other, called the cathode, has a negative charge. In the tube that Crookes used, the metal cathode was a disk at one end of the tube. In the center of the tube was an object shaped like a cross, as you can see in **Figure 4.** When the battery was connected, the glass tube suddenly lit up with a greenish-colored glow. A shadow of the object appeared at the opposite end of the tube—the anode. The shadow showed Crookes that something was traveling in a straight line from the cathode to the anode, similar to the beam of a flashlight. The cross-shaped object was getting in the way of the beam and blocking it, just like when a road crew uses a stencil to block paint from certain places on the road when they are marking lanes and arrows. You can see this in **Figure 5.**

Figure 5 Paint passing by a stencil is an example of what happened with Crookes' tube, the cathode ray, and the cross.

Cathode Rays Crookes hypothesized that the green glow in the tube was caused by rays, or streams of particles. These rays were called cathode rays because they were produced at the cathode. Crookes' tube is known as a cathode-ray tube, or CRT. **Figure 6** shows a CRT. They were used for TV and computer display screens for many years now.

☑ **Reading Check**

What are cathode rays?

Discovering Charged Particles

The news of Crookes' experiments excited the scientific community of the time. But many scientists were not convinced that the cathode rays were streams of particles. Was the greenish glow light, or was it a stream of charged particles? In 1897, J.J. Thomson, an English physicist, tried to clear up the confusion. He placed a magnet beside the tube from Crookes' experiments. In **Figure 7,** you can see that the beam is bent in the direction of the magnet. Light cannot be bent by a magnet, so the beam couldn't be light. Therefore, Thomson concluded that the beam must be made up of charged particles of matter that came from the cathode.

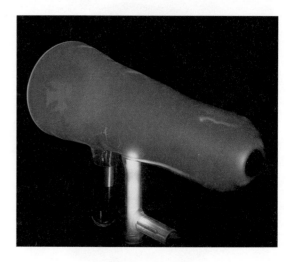

Figure 6 The cathode-ray tube got its name because the particles start at the cathode and travel to the anode. At one time, a CRT was in every TV and computer monitor.

The Electron Thomson then repeated the CRT experiment using different metals for the cathode and different gases in the tube. He found that the same charged particles were produced no matter what elements were used for the cathode or the gas in the tube. Thomson concluded that cathode rays are negatively charged particles of matter. How did Thomson know the particles were negatively charged? He knew that opposite charges attract each other. He observed that these particles were attracted to the positively charged anode, so he reasoned that the particles must be negatively charged.

These negatively charged particles are now called **electrons.** Thomson also inferred that electrons are a part of every kind of atom because they are produced by every kind of cathode material. Perhaps the biggest surprise that came from Thomson's experiments was the evidence that particles smaller than the atom do exist.

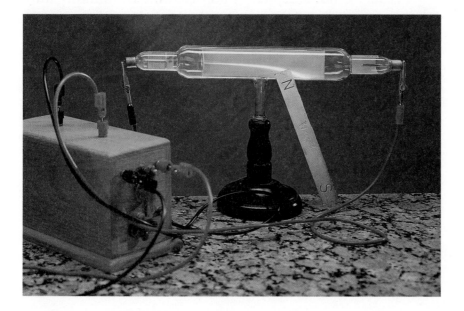

Figure 7 When a magnet was placed near a CRT, the cathode rays were bent. Since light is not bent by a magnet, Thomson determined that cathode rays were made of charged particles.

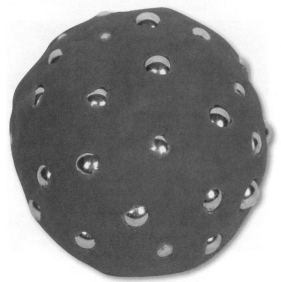

Figure 8 Modeling clay with ball bearings mixed through is another way to picture the J.J. Thomson atom. The clay contains all the positive charge of the atom. The ball bearings, which represent the negatively charged electrons, are mixed evenly in the clay. **Explain** *why Thomson included positive particles in his atomic model.*

Thomson's Atomic Model Some of the questions posed by scientists were answered in light of Thomson's experiments. However, the answers inspired new questions. If atoms contain one or more negatively charged particles, then all matter, which is made of atoms, should be negatively charged as well. But all matter isn't negatively charged. Could it be that atoms also contain some positive charge? The negatively charged electrons and the unknown positive charge would then neutralize each other in the atom. Thomson came to this conclusion and included positive charge in his model of the atom.

Using his new findings, Thomson revised Dalton's model of the atom. Instead of a solid ball that was the same throughout, Thomson pictured a sphere of positive charge. The negatively charged electrons were spread evenly among the positive charge. This is modeled by the ball of clay shown in **Figure 8.** The positive charge of the clay is equal to the negative charge of the electrons. Therefore, the atom is neutral. It was later discovered that not all atoms are neutral. The number of electrons within an element can vary. If there is more positive charge than negative electrons, the atom has an overall positive charge. If there are more negative electrons than positive charge, the atom has an overall negative charge.

 **Reading Check** *What particle did Thomson's model have scattered through it?*

Rutherford's Experiments

A model is not accepted in the scientific community until it has been tested and the tests support previous observations. In 1906, Ernest Rutherford and his coworkers began an experiment to find out if Thomson's model of the atom was correct. They wanted to see what would happen when they fired fast-moving, positively charged bits of matter, called alpha particles, at a thin film of a metal such as gold. Alpha particles, which come from unstable atoms, are positively charged, and so they are repelled by particles of matter which also have a positive charge.

Figure 9 shows how the experiment was set up. A source of alpha particles was aimed at a thin sheet of gold foil that was only 400 nm thick. The foil was surrounded by a fluorescent (floo REH sunt) screen that gave a flash of light each time it was hit by a charged particle.

Expected Results Rutherford was certain he knew what the results of this experiment would be. His prediction was that most of the speeding alpha particles would pass right through the foil and hit the screen on the other side, just like a bullet fired through a pane of glass. Rutherford reasoned that the thin, gold film did not contain enough matter to stop the speeding alpha particle or change its path. Also, there wasn't enough charge in any one place in Thomson's model to repel the alpha particle strongly. He thought that the positive charge in the gold atoms might cause a few minor changes in the path of the alpha particles. However, he assumed that this would only occur a few times.

That was a reasonable hypothesis because in Thomson's model, the positive charge is essentially neutralized by nearby electrons. Rutherford was so sure of what the results would be that he turned the work over to a graduate student.

The Model Fails Rutherford was shocked when his student rushed in to tell him that some alpha particles were veering off at large angles. You can see this in **Figure 9.** Rutherford expressed his amazement by saying, "It was about as believable as if you had fired a 15-inch shell at a piece of tissue paper, and it came back and hit you." How could such an event be explained? The positively charged alpha particles were moving with such high speed that it would take a large positive charge to cause them to bounce back. The uniform mix of mass and charges in Thomson's model of the atom did not allow for this kind of result.

Figure 9 In Rutherford's experiment, alpha particles bombarded the gold foil. Most particles passed right through the foil or veered slightly from a straight path, but some particles bounced right back. The path of a particle is shown by a flash of light when it hits the fluorescent screen.

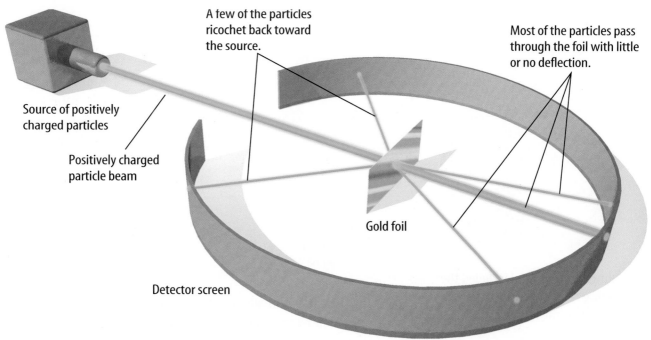

A few of the particles ricochet back toward the source.

Most of the particles pass through the foil with little or no deflection.

Source of positively charged particles

Positively charged particle beam

Gold foil

Detector screen

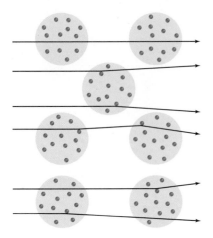

• Proton   ⟶ Path of alpha particle

Figure 10 Rutherford thought that if the atom could be described by Thomson's model, as shown above, then only minor bends in the paths of the particles would have occurred.

A Model with a Nucleus

Now Rutherford and his team had to come up with an explanation for these unexpected results. They might have drawn diagrams like those in **Figure 10,** which uses Thomson's model and shows what Rutherford expected. Now and then, an alpha particle might be affected slightly by a positive charge in the atom and turn a bit off course. However, large changes in direction were not expected.

The Proton The actual results did not fit this model, so Rutherford proposed a new one, shown in **Figure 11.** He hypothesized that almost all the mass of the atom and all of its positive charge are crammed into an incredibly small region of space at the center of the atom called the nucleus. Eventually, his prediction was proved true. In 1920 scientists identified the positive charges in the nucleus as protons. A **proton** is a positively charged particle present in the nucleus of all atoms. The rest of each atom is empty space occupied by the atom's almost-massless electrons.

✔️ **Reading Check** *How did Rutherford describe his new model?*

Figure 12 shows how Rutherford's new model of the atom fits the experimental data. Most alpha particles could move through the foil with little or no interference because of the empty space that makes up most of the atom. However, if an alpha particle made a direct hit on the nucleus of a gold atom, which has 79 protons, the alpha particle would be strongly repelled and bounce back.

Figure 11 The nuclear model was new and helped explain experimental results.

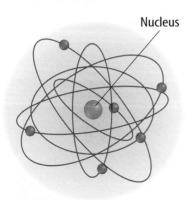

Nucleus

Rutherford's model included the dense center of positive charge known as the nucleus.

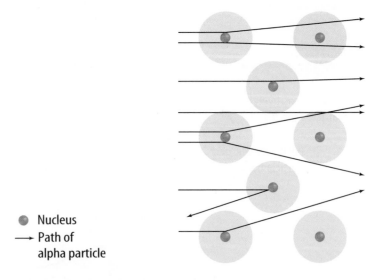

● Nucleus
⟶ Path of alpha particle

Figure 12 This nucleus that contained most of the mass of the atom caused the deflections that were observed in his experiment.

The Neutron Rutherford's nuclear model was applauded as other scientists reviewed the results of the experiments. However, some data didn't fit. Once again, more questions arose and the scientific process continued. For instance, an atom's electrons have almost no mass. According to Rutherford's model, the only other particle in the atom was the proton. That meant that the mass of an atom should have been approximately equal to the mass of its protons. However, it wasn't. The mass of most atoms is at least twice as great as the mass of its protons. That left scientists with a dilemma and raised a new question. Where does the extra mass come from if only protons and electrons make up the atom?

It was proposed that another particle must be in the nucleus to account for the extra mass. The particle, which was later called the **neutron** (NEW trahn), would have the same mass as a proton and be electrically neutral. Proving the existence of neutrons was difficult though, because a neutron has no charge. Therefore, the neutron doesn't respond to magnets or cause fluorescent screens to light up. It took another 20 years before scientists were able to show by more modern experiments that atoms contain neutrons.

Reading Check *What particles are in the nucleus of the nuclear atom?*

The model of the atom was revised again to include the newly discovered neutrons in the nucleus. The nuclear atom, shown in **Figure 13,** has a tiny nucleus tightly packed with positively charged protons and neutral neutrons. Negatively charged electrons occupy the space surrounding the nucleus. The number of electrons in a neutral atom equals the number of protons in the atom.

Mini LAB

Modeling the Nuclear Atom

Procedure
1. On a sheet of **paper,** draw a circle with a diameter equal to the width of the paper.
2. **Small dots of paper in two colors** will represent protons and neutrons. Using a dab of **glue** on each paper dot, make a model of the nucleus of the oxygen atom in the center of your circle. Oxygen has eight protons and eight neutrons.

Analysis
1. What particle is missing from your model of the oxygen atom?
2. How many of that missing particle should there be, and where should they be placed?

Try at Home

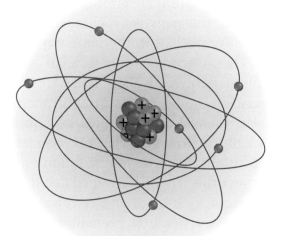

Figure 13 This atom of carbon, atomic number 6, has six protons and six neutrons in its nucleus.
Identify *how many electrons are in the "empty" space surrounding the nucleus.*

Figure 14 If this Ferris wheel in London, with a diameter of 132 m, were the outer edge of the atom, the nucleus would be about the size of a single letter *o* on this page.

Size and Scale Drawings of the nuclear atom such as the one in **Figure 13** don't give an accurate representation of the extreme smallness of the nucleus compared to the rest of the atom. For example, if the nucleus were the size of a table-tennis ball, the atom would have a diameter of more than 2.4 km. Another way to compare the size of a nucleus with the size of the atom is shown in **Figure 14.** Perhaps now you can see better why in Rutherford's experiment, most of the alpha particles went directly through the gold foil without any interference from the gold atoms. Plenty of empty space allows the alpha particles an open pathway.

Further Developments

Even into the twentieth century, physicists were working on a theory to explain how electrons are arranged in an atom. It was natural to think that the negatively charged electrons are attracted to the positive nucleus in the same way the Moon is attracted to Earth. Then, electrons would travel in orbits around the nucleus. A physicist named Niels Bohr even calculated exactly what energy levels those orbits would represent for the hydrogen atom. His calculations explained experimental data found by other scientists. However, scientists soon learned that electrons are in constant, unpredictable motion and can't be described easily by an orbit. They determined that it was impossible to know the precise location of an electron at any particular moment. Their work inspired even more research and brainstorming among scientists around the world.

**INTEGRATE
Physics**

Physicists In the 1920s, physicists began to think that electrons—like light—have a wave/particle nature. This is called quantum theory. Research which two scientists introduced this theory. In your Science Journal, infer how thoughts about atoms changed.

Electrons as Waves Physicists began to wrestle with explaining the unpredictable nature of electrons. Surely the experimental results they were seeing and the behavior of electrons could somehow be explained with new theories and models. The unconventional solution was to understand electrons not as particles, but as waves. This led to further mathematical models and equations that brought much of the experimental data together.

The Electron Cloud Model The new model of the atom allows for the somewhat unpredictable wave nature of electrons by defining a region where the electron is most likely to be found. Electrons travel in a region surrounding the nucleus, which is called the **electron cloud.** The current model for the electron cloud is shown in **Figure 15.** The electrons are more likely to be close to the nucleus rather than farther away because they are attracted to the positive charges of the protons. Notice the fuzzy outline of the cloud. Because the electrons could be anywhere, the cloud has no firm boundary. Interestingly, within the electron cloud, the electron in a hydrogen atom probably is found in the region Bohr calculated.

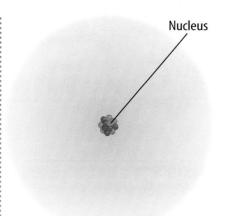

Nucleus

Figure 15 The electrons are more likely to be close to the nucleus rather than farther away, but they could be anywhere. **Explain** *why the electrons woud be closer to the nucleus.*

section 1 review

Summary

Models of the Atom
- Some early philosophers believed all matter was made of small particles.
- John Dalton proposed that all matter is made of atoms that were hard spheres.
- J. J. Thomson showed that the particles in a CRT were negatively charged particles, later called electrons. These were smaller than an atom. He proposed the atom as a sphere of positive charge with electrons spread evenly among the charge.
- In his experiments, Rutherford showed that positive charge existed in a small region of the atom which he called the nucleus. The positive charge was called a proton.
- In order to explain the mass of an atom, the neutron was proposed, an uncharged particle the same mass as a proton and in the nucleus.
- Electrons are now believed to move about the nucleus in an electron cloud.

Self Check

1. **Explain** how the nuclear atom differs from the uniform sphere model of the atom.
2. **Determine** how many electrons a neutral atom with 49 protons has.
3. **Describe** what cathode rays are and how they were discovered.
4. **Think Critically** In Rutherford's experiment, why wouldn't the electrons in the atoms of the gold foil affect the paths of the alpha particles.
5. **Concept Map** Design and complete a concept map using all the words in the vocabulary list for this section. Add any other terms or words that will help create a complete diagram of the section and the concepts in contains.

Applying Math

6. **Solve One-Step Equations** The mass of an electron is 9.11×10^{-28} g. The mass of a proton is 1,836 times more than that of the electron. Calculate the mass of the proton in grams and convert that mass into kilograms.

The Simplest Matter

What You'll Learn

- **Describe** the relationship between elements and the periodic table.
- **Explain** the meaning of atomic mass and atomic number.
- **Identify** what makes an isotope.
- **Contrast** metals, metalloids, and nonmetals.

Why It's Important

Everything on Earth is made of the elements that are listed on the periodic table.

Review Vocabulary

mass: a measure of the amount of matter

New Vocabulary

- atomic number
- isotope
- mass number
- atomic mass
- metals
- nonmetals
- metalloids

The Elements

Have you watched television today? TV sets are common, yet each one is a complex system. The outer case is made mostly of plastic, and the screen is made of glass. Many of the parts that conduct electricity are metals or combinations of metals. Other parts in the interior of the set contain materials that barely conduct electricity. All of the different materials have one thing in common. They are made up of even simpler materials. In fact, if you had the proper equipment, you could separate the plastics, glass, and metals into these simpler materials.

One Kind of Atom Eventually, though, you would separate the materials into groups of atoms. At that point, you would have a collection of elements. Recall that an element is matter made of only one kind of atom. At least 115 elements are known and about 90 of them occur naturally on Earth. These elements make up gases in the air, minerals in rocks, and liquids such as water. Examples of naturally occurring elements include the oxygen and nitrogen in the air you breathe and the metals gold, silver, aluminum, and iron. The other elements are known as synthetic elements. These elements have been made in nuclear reactions by scientists with machines called particle accelerators, like the one shown in **Figure 16.** Some synthetic elements have important uses in medical testing and are found in smoke detectors and heart pacemaker batteries.

Figure 16 The Tevatron has a circumference of 6.3 km—a distance that allows particles to accelerate to high speeds. These high-speed collisions can create synthetic elements.

Figure 17 When you look for information in the library, a system of organization called the Dewey Decimal Classification System helps you find a book quickly and efficiently.

Dewey Decimal Classification System	
000	Computers, information, and general reference
100	Philosophy and psychology
200	Religion
300	Social sciences
400	Language
500	Science
600	Technology
700	Arts and recreation
800	Literature
900	Philosophy and psychology

The Periodic Table

Suppose you go to a library, like the one shown in **Figure 17,** to look up information for a school assignment. How would you find the information? You could look randomly on shelves as you walk up and down rows of books, but the chances of finding your book would be slim. Not only that, you also would probably become frustrated in the process. To avoid such haphazard searching, some libraries use the Dewey Decimal Classification System to categorize and organize their volumes and to help you find books quickly and efficiently.

Charting the Elements When scientists need to look up information about an element or select one to use in the laboratory, they need to be quick and efficient, too. Chemists have created a chart called the periodic table of the elements to help them organize and display the elements. **Figure 18** shows how scientists changed their model of the periodic table over time.

On the inside back cover of this book, you will find a modern version of the periodic table. Each element is represented by a chemical symbol that contains one to three letters. The symbols are a form of chemical shorthand that chemists use to save time and space—on the periodic table as well as in written formulas. The symbols are an important part of an international system that is understood by scientists everywhere.

The elements are organized on the periodic table by their properties. There are rows and columns that represent relationships between the elements. The rows in the table are called periods. The elements in a row have the same number of energy levels. The columns are called groups. The elements in each group have similar properties related to their structure. They also tend to form similar bonds.

Dewey Decimal System Melvil Dewey is the man responsible for organizing our knowledge and libraries. His working in the Amherst College library led him to propose a method of classifying books. The Dewey Decimal System divides books into ten categories. Since 1876, this classification system has helped us locate information easily.

Figure 18

The familiar periodic table that adorns many science classrooms is based on a number of earlier efforts to identify and classify the elements. In the 1790s, one of the first lists of elements and their compounds was compiled by French chemist Antoine-Laurent Lavoisier, who is shown in the background picture with his wife and assistant, Marie Anne. Three other tables are shown here.

John Dalton (Britain, 1803) used symbols to represent elements. His table also assigned masses to each element.

An early alchemist put together this table of elements and compounds. Some of the symbols have their origin in astrology.

Dmitri Mendeleev (Russia, 1869) arranged the 63 elements known to exist at that time into groups based on their chemical properties and atomic weights. He left gaps for elements he predicted were yet to be discovered.

Identifying Characteristics

Each element is different and has unique properties. These differences can be described in part by looking at the relationships between the atomic particles in each element. The periodic table contains numbers that describe these relationships.

Number of Protons and Neutrons Look up the element chlorine on the periodic table found on the inside back cover of your book. Cl is the symbol for chlorine, as shown in **Figure 19,** but what are the two numbers? The top number is the element's **atomic number.** It tells you the number of protons in the nucleus of each atom of that element. Every atom of chlorine, for example, has 17 protons in its nucleus.

 **Reading Check** *What are the atomic numbers for Cs, Ne, Pb, and U?*

Isotopes Although the number of protons changes from element to element, every atom of the same element has the same number of protons. However, the number of neutrons can vary even for one element. For example, some chlorine atoms have 18 neutrons in their nucleus while others have 20. These two types of chlorine atoms are chlorine-35 and chlorine-37. They are called **isotopes** (I suh tohps), which are atoms of the same element that have different numbers of neutrons.

You can tell someone exactly which isotope you are referring to by using its mass number. An atom's **mass number** is the number of protons plus the number of neutrons it contains. The numbers 35 and 37, which were used to refer to chlorine, are mass numbers. Hydrogen has three isotopes with mass numbers of 1, 2, and 3. They are shown in **Figure 20.** Each hydrogen atom always has one proton, but in each isotope the number of neutrons is different.

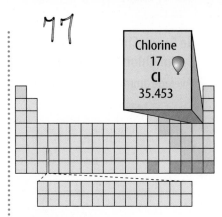

Figure 19 The periodic table block for chlorine shows its symbol, atomic number, and atomic mass.
Determine *if chlorine atoms are more or less massive than carbon atoms.*

Figure 20 Three isotopes of hydrogen are known to exist. They have zero, one, and two neutrons in addition to their one proton. Protium, with only the one proton, is the most abundant isotope.

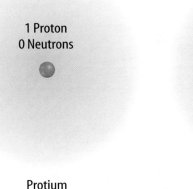

1 Proton
0 Neutrons

Protium

1 Proton
1 Neutron

Deuterium

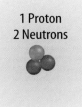

1 Proton
2 Neutrons

Tritium

Circle Graph Showing Abundance of Chlorine Isotopes
Average atomic mass = 35.45 u

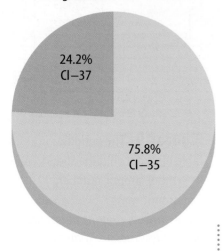

Figure 21 If you have 1,000 atoms of chlorine, about 758 will be chlorine-35 and have a mass of 34.97 u each. About 242 will be chlorine-37 and have a mass of 36.97 u each. The total mass of the 1,000 atoms is 35,454 u, so the average mass of one chlorine atom is about 35.45 u.

Atomic Mass The **atomic mass** is the weighted average mass of the isotopes of an element. The atomic mass is the number found below the element symbol in **Figure 19.** The unit that scientists use for atomic mass is called the atomic mass unit, which is given the symbol u. It is defined as 1/12 the mass of a carbon-12 atom.

The calculation of atomic mass takes into account the different isotopes of the element. Chlorine's atomic mass of 35.45 u could be confusing because there aren't any chlorine atoms that have that exact mass. About 76 percent of chlorine atoms are chlorine-35 and about 24 percent are chlorine-37, as shown in **Figure 21.** The weighted average mass of all chlorine atoms is 35.45 u.

Classification of Elements

Elements fall into three general categories—metals, metalloids (ME tuh loydz), and nonmetals. The elements in each category have similar properties.

Metals generally have a shiny or metallic luster and are good conductors of heat and electricity. All metals, except mercury, are solids at room temperature. Metals are malleable (MAL yuh bul), which means they can be bent and pounded into various shapes. The beautiful form of the shell-shaped basin in **Figure 22** is a result of this characteristic. Metals are also ductile, which means they can be drawn into wires without breaking. If you look at the periodic table, you can see that most of the elements are metals.

Figure 22 The artisan is chasing, or chiseling, the malleable metal into the desired form.

Other Elements **Nonmetals** are elements that are usually dull in appearance. Most are poor conductors of heat and electricity. Many are gases at room temperature, and bromine is a liquid. The solid nonmetals are generally brittle, meaning they cannot change shape easily without breaking. The nonmetals are essential to the chemicals of life. More than 97 percent of your body is made up of various nonmetals, as shown in **Figure 23.** You can see that, except for hydrogen, the nonmetals are found on the right side of the periodic table.

Metalloids are elements that have characteristics of metals and nonmetals. On the periodic table, metalloids are found between the metals and nonmetals. All metalloids are solids at room temperature. Some metalloids are shiny and many are conductors, but they are not as good at conducting heat and electricity as metals are. Some metalloids, such as silicon, are used to make the electronic circuits in computers, televisions, and other electronic devices.

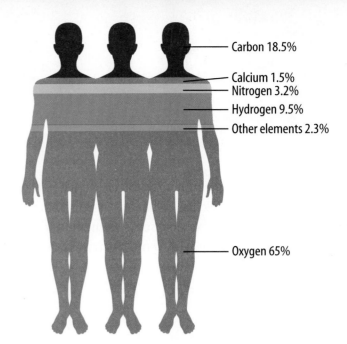

Carbon 18.5%

Calcium 1.5%
Nitrogen 3.2%
Hydrogen 9.5%
Other elements 2.3%

Oxygen 65%

Figure 23 You are made up of mostly nonmetals.

section 2 review

Summary

The Elements

- An element is matter made of only one type of atom.
- Some elements occur naturally on Earth. Synthetic elements are made in nuclear reactions in particle accelerators.

The Periodic Table

- The periodic table arranges and displays all known elements in an orderly way.
- Each element has been given a chemical symbol that is used on a periodic table.

Identifying Characteristics

- Each element has a unique number of protons, called the atomic mass number.
- Isotopes of elements are important when determining the atomic mass of an element.

Classification of Elements

- Elements are divided into three categories based on certain properties: metal, metalloids, and nonmetals.

Self Check

1. **Explain** some of the uses of metals based on their properties.
2. **Describe** the difference between atomic number and atomic mass.
3. **Define** the term *isotope.* Explain how two isotopes of an element are different.
4. **Think Critically** Describe how to find the atomic number for the element oxygen. Explain what this information tells you about oxygen.
5. **Interpret Data** Look up the atomic mass of the element boron in the periodic table inside the back cover of this book. The naturally occurring isotopes of boron are boron-10 and boron-11. Explain which of the two isotopes is more abundant?

Applying Math

6. **Solve One-Step Equations** An atom of niobium has a mass number of 93. How many neutrons are in the nucleus of this atom? An atom of phosphorus has 15 protons and 15 neutrons in the nucleus. What is the mass number of this isotope?

Elements and the Periodic Table

The periodic table organizes the elements, but what do they look like? What are they used for? In this lab, you'll examine some elements and share your findings with your classmates.

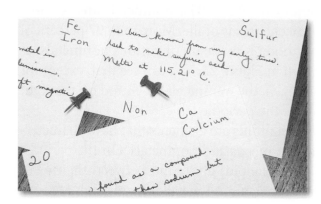

● Real-World Questions

What are some of the characteristics and purposes of the chemical elements?

Goals

- **Classify** the chemical elements.
- **Organize** the elements into the groups and periods of the periodic table.

Materials

colored markers	large bulletin board
large index cards	81/2-in × 14-in paper
Merck Index	thumbtacks
encyclopedia	*pushpins
*other reference materials	*Alternate materials

Safety Precautions

WARNING: *Use care when handling sharp objects.*

● Procedure

1. Select the assigned number of elements from the list provided by your teacher.

2. **Design** an index card for each of your selected elements. On each card, mark the element's atomic number in the upper left-hand corner and write its symbol and name in the upper right-hand corner.

3. **Research** each of the elements and write several sentences on the card about its appearance, its other properties, and its uses.

4. **Classify** each element as a metal, a metalloid, or a nonmetal based upon its properties.

5. **Write** the appropriate classification on each of your cards using the colored marker chosen by your teacher.

6. Work with your classmates to make a large periodic table. Use thumbtacks to attach your cards to a bulletin board in their proper positions on the periodic table.

7. **Draw** your own periodic table. Place the elements' symbols and atomic numbers in the proper locations on your table.

● Conclude and Apply

1. **Interpret** the class data and classify the elements into the categories metal, metalloid, and nonmetal. Highlight each category in a different color on your periodic table.

2. **Predict** the properties of a yet-undiscovered element located directly under francium on the periodic table.

Communicating Your Data

Compare and contrast your table with that of a friend. Discuss the differences. **For more help, refer to the** Science Skill Handbook.

Compounds and Mixtures

Substances

Scientists classify matter in several ways that depend on what it is made of and how it behaves. For example, matter that has the same composition and properties throughout is called a **substance.** Elements, such as a bar of gold or a sheet of aluminum, are substances. When different elements combine, other substances are formed.

Compounds The elements hydrogen and oxygen exist as separate, colorless gases. However, these two elements can combine, as shown in **Figure 24,** to form the compound water, which is different from the elements that make it up. A **compound** is a substance whose smallest unit is made up of atoms of more than one element bonded together.

Compounds often have properties that are different from the elements that make them up. Water is distinctly different from the elements that make it up. It is also different from another compound made from the same elements. Have you ever used hydrogen peroxide (H_2O_2) to disinfect a cut? This compound is a different combination of hydrogen and oxygen and has different properties from those of water.

Water is a nonirritating liquid that is used for bathing, drinking, cooking, and much more. In contrast, hydrogen peroxide carries warnings on its labels such as *Keep Hydrogen Peroxide Out of the Eyes.* Although it is useful in solutions for cleaning contact lenses, it is not safe for your eyes as it comes from the bottle.

as you read

What **You'll Learn**

■ **Identify** the characteristics of a compound.
■ **Compare and contrast** different types of mixtures.

Why **It's Important**

The food you eat, the materials you use, and all matter can be classified by compounds or mixtures.

🔍 **Review Vocabulary**
formula: shows which elements and how many atoms of each make up a compound.

New Vocabulary
● substance
● compound
● mixture

Figure 24 A space shuttle is powered by the reaction between liquid hydrogen and liquid oxygen. The reaction produces a large amount of energy and the compound water. **Explain** *why a car that burns hydrogen rather than gasoline would be friendly to the environment.*

Figure 25 The elements hydrogen and oxygen can form two compounds—water and hydrogen peroxide. Note the differences in their structure.

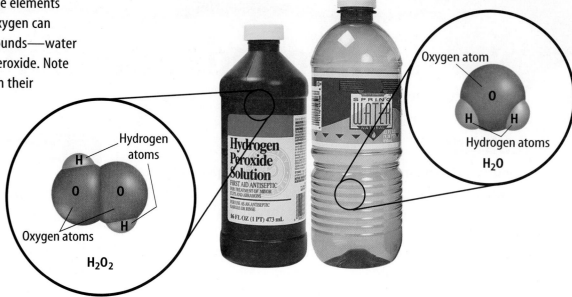

H₂O₂

Hydrogen atoms

Oxygen atoms

Oxygen atom

Hydrogen atoms

H₂O

Mini LAB

Comparing Compounds

Procedure

1. Collect the following substances—**granular sugar, rubbing alcohol,** and **salad oil.**
2. Observe the color, appearance, and state of each substance. Note the thickness or texture of each substance.
3. Stir a spoonful of each substance into separate **beakers of hot water** and observe.

Analysis

1. Compare the different properties of the substances.
2. The formulas of the three substances are made of only carbon, hydrogen, and oxygen. Infer how they can have different properties.

Compounds Have Formulas What's the difference between water and hydrogen peroxide? H_2O is the chemical formula for water, and H_2O_2 is the formula for hydrogen peroxide. The formula tells you which elements make up a compound as well as how many atoms of each element are present. Look at **Figure 25.** The subscript number written below and to the right of each element's symbol tells you how many atoms of that element exist in one unit of that compound. For example, hydrogen peroxide has two atoms of hydrogen and two atoms of oxygen. Water is made up of two atoms of hydrogen and one atom of oxygen.

Carbon dioxide, CO_2, is another common compound. Carbon dioxide is made up of one atom of carbon and two atoms of oxygen. Carbon and oxygen also can form the compound carbon monoxide, CO, which is a gas that is poisonous to all warm-blooded animals. As you can see, no subscript is used when only one atom of an element is present. A given compound always is made of the same elements in the same proportion. For example, water always has two hydrogen atoms for every oxygen atom, no matter what the source of the water is. No matter what quantity of the compound you have, the formula of the compound always remains the same. If you have 12 atoms of hydrogen and six atoms of oxygen, the compound is still written H_2O, but you have six molecules of H_2O ($6 H_2O$), not $H_{12}O_6$. The formula of a compound communicates its identity and makeup to any scientist in the world.

✔ **Reading Check** *Propane has three carbon and eight hydrogen atoms. What is its chemical formula?*

Mixtures

When two or more substances (elements or compounds) come together but don't combine to make a new substance, a **mixture** results. Unlike compounds, the proportions of the substances in a mixture can be changed without changing the identity of the mixture. For example, if you put some sand into a bucket of water, you have a mixture of sand and water. If you add more sand or more water, it's still a mixture of sand and water. Its identity has not changed. Air is another mixture. Air is a mixture of nitrogen, oxygen, and other gases, which can vary at different times and places. Whatever the proportion of gases, it is still air. Even your blood is a mixture that can be separated, as shown in **Figure 26** by a machine called a centrifuge.

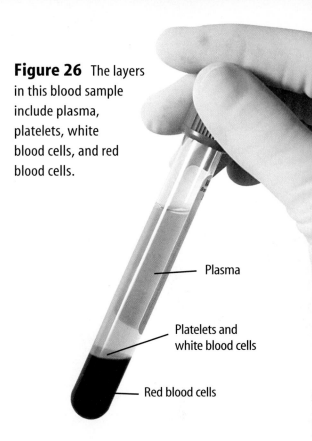

Figure 26 The layers in this blood sample include plasma, platelets, white blood cells, and red blood cells.

Plasma

Platelets and white blood cells

Red blood cells

 **Reading Check** *How do the proportions of a mixture relate to its identity?*

Applying Science

What's the best way to desalt ocean water?

You can't drink ocean water because it contains salt and other suspended materials. Or can you? In many areas of the world where drinking water is in short supply, methods for getting the salt out of salt water are being used to meet the demand for fresh water. Use your problem solving skills to find the best method to use in a particular area.

Methods for Desalting Ocean Water			
Process	Amount of Water a Unit Can Desalt in a Day (m³)	Special Needs	Number of People Needed to Operate
Distillation	1,000 to 200,000	lots of energy to boil the water	many
Electrodialysis	10 to 4,000	stable source of electricity	1 to 2 persons

Identifying the Problem

The table above compares desalting methods. In distillation, the ocean water is heated. Pure water boils off and is collected, and the salt is left behind. Electrodialysis uses electric current to pull salt particles out of water.

Solving the Problem

1. What method(s) might you use to desalt the water for a large population where energy is plentiful?
2. What method(s) would you choose to use in a single home?

Figure 27 Mixtures are part of your everyday life.

Science online

Topic: Mixtures

Visit bookk.msscience.com for Web links to information about separating mixtures.

Activity Describe how chemists separate the components of a mixture.

INTEGRATE Life Science Your blood is a mixture made up of elements and compounds. It contains white blood cells, red blood cells, water, and a number of dissolved substances. The different parts of blood can be separated and used by doctors in different ways. The proportions of the substances in your blood change daily, but the mixture does not change its identity.

Separating Mixtures Sometimes you can use a liquid to separate a mixture of solids. For example, if you add water to a mixture of sugar and sand, only the sugar dissolves in the water. The sand then can be separated from the sugar and water by pouring the mixture through a filter. Heating the remaining solution will separate the water from the sugar.

At other times, separating a mixture of solids of different sizes might be as easy as pouring them through successively smaller sieves or filters. A mixture of marbles, pebbles, and sand could be separated in this way.

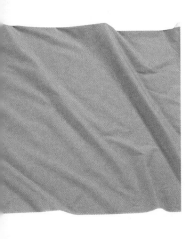

Homogeneous or Heterogeneous

Mixtures, such as the ones shown in **Figure 27,** can be classified as homogeneous or heterogeneous. *Homogeneous* means "the same throughout." You can't see the different parts in this type of mixture. In fact, you might not always know that homogeneous mixtures are mixtures because you can't tell by looking. Which mixtures in **Figure 27** are homogeneous? No matter how closely you look, you can't see the individual parts that make up air or the parts of the mixture called brass in the lamp shown. Homogeneous mixtures can be solids, liquids, or gases.

A heterogeneous mixture has larger parts that are different from each other. You can see the different parts of a heterogeneous mixture, such as sand and water. How many heterogeneous mixtures are in **Figure 27?** A pepperoni and mushroom pizza is a tasty kind of heterogeneous mixture. Other examples of this kind of mixture include tacos, vegetable soup, a toy box full of toys, or a tool box full of nuts and bolts.

INTEGRATE Earth Science

Rocks and Minerals
Scientists called geologists study rocks and minerals. A mineral is composed of a pure substance. Rocks are mixtures and can be described as being homogeneous or heterogeneous. Research to learn more about rocks and minerals and note some examples of homogeneous and heterogeneous rocks in your Science Journal.

section 3 review

Summary

Substances

- A substance can be either an element or a compound.
- A compound contains more than one kind of element bonded together.
- A chemical formula shows which elements and how many atoms of each make up a compound.

Mixtures

- A mixture contains substances that are not chemically bonded together.
- There are many ways to separate mixtures based on their physical properties.
- Homogeneous mixtures are those that are the same throughout. These types of mixtures can be solids, liquids, or gases.
- Heterogeneous mixtures have larger parts that are different from each other.

Self Check

1. **List** three examples of compounds and three examples of mixtures. Explain your choices.
2. **Describe** a procedure that can be used to separate a liquid homogenous mixture of salt and water.
3. **Identify** the elements that make up the following compounds: H_2SO_4 and $CHCl_3$.
4. **Think Critically** Explain whether your breakfast was a compound, a homogeneous mixture, or a heterogeneous mixture.

Applying Skills

5. **Compare and contrast** compounds and mixtures based on what you have learned from this section.
6. **Use a Database** Use a computerized card catalog or database to find information about one element from the periodic table. Include information about the properties and the uses of the mixtures and/or compounds in which the element is frequently found.

Mystery Mixture

Goals

- **Test** for the presence of certain compounds.
- **Decide** which of these compounds are present in an unknown mixture.

Materials

test tubes (4)
cornstarch
powdered sugar
baking soda
mystery mixture
small scoops (3)
dropper bottles (2)
iodine solution
white vinegar
hot plate
250-mL beaker
water (125 mL)
test-tube holder
small pie pan

Safety Precautions

WARNING: *Use caution when handling hot objects. Substances could stain or burn clothing. Be sure to point the test tube away from your face and your classmates while heating.*

⊙ *Real-World Question*

You will encounter many compounds that look alike. For example, a laboratory stockroom is filled with white powders. It is important to know what each is. In a kitchen, cornstarch, baking powder, and powdered sugar are compounds that look alike. To avoid mistaking one for another, you can learn how to identify them. Different compounds can be identified

by using chemical tests. For example, some compounds react with certain liquids to produce gases. Other combinations produce distinctive colors. Some compounds have high melting points. Others have low melting points. How can the compounds in an unknown mixture be identified by experimentation?

◉ Procedure

1. Copy the data table into your Science Journal. Record your results carefully for each of the following steps.

2. Again place a small scoopful of cornstarch on the pie pan. Do the same for the sugar and baking soda maintaining separate piles. Add a drop of vinegar to each. Wash and dry the pan after you record your observations.

3. Again place a small scoopful of cornstarch, sugar, and baking soda on the pie pan. Add a drop of iodine solution to each one.

Identifying Presence of Compounds

Substance to Be Tested	Fizzes with Vinegar	Turns Blue with Iodine	Melts When Heated
Cornstarch			
Sugar	Do not write in this book.		
Baking soda			
Mystery mix			

4. Place a small scoopful of each compound in a separate test tube. Hold the test tube with the test-tube holder and with an oven mitt. Gently heat the test tube in a beaker of boiling water on a hot plate.

5. Follow steps 2 through 4 to test your mystery mixture for each compound.

◉ Analyze Your Data

1. **Identify** from your data table which compound(s) you have.

2. **Describe** how you decided which substances were in your unknown mixture.

◉ Conclude and Apply

1. **Explain** how you would be able to tell if all three compounds were not in your mystery substance.

2. **Draw a Conclusion** What would you conclude if you tested baking powder from your kitchen and found that it fizzed with vinegar, turned blue with iodine, and did not melt when heated?

𝒞ommunicating Your Data

Make a different data table to display your results in a new way. **For more help, refer to the** Science Skill Handbook.

Ancient Views of Matter

Two cultures observed the world around them differently

Water

Air & ether

The world's earliest scientists were people who were curious about the world around them and who tried to develop explanations for the things they observed. This type of observation and inquiry flourished in ancient cultures such as those found in India and China. Read on to see how the ancient Indians and Chinese defined matter.

Indian Ideas

To Indians living about 3,000 years ago, the world was made up of five elements: fire, air, earth, water, and ether, which they thought of as an unseen substance that filled the heavens. Building upon this concept, the early Indian philosopher Kashyapa (kah SHI ah pah) proposed that the five elements could be broken down into smaller units called parmanu (par MAH new). Parmanu were similar to atoms in that they were too small to be seen but still retained the properties of the original element. Kashyapa also believed that each type of parmanu had unique physical and chemical properties.

Metal

Parmanu of earth elements, for instance, were heavier than parmanu of air elements. The different properties of the parmanu determined the characteristics of a substance. Kashyapa's ideas about matter are similar to those of the Greek philosopher Democritus, who lived centuries after Kashyapa.

Chinese Ideas

The ancient Chinese also broke matter down into five elements: fire, wood, metal, earth, and water. Unlike the early Indians, however, the Chinese believed that the elements constantly changed form. For example, wood can be burned and thus changes to fire. Fire eventually dies down and becomes ashes, or earth. Earth gives forth metals from the ground. Dew or water collects on these metals, and the water then nurtures plants that grow into trees, or wood.

This cycle of constant change was explained in the fourth century B.C. by the philosopher Tsou Yen. Yen, who is known as the founder of Chinese scientific thought, wrote that all changes that took place in nature were linked to changes in the five elements.

Fire

Earth

Research Write a brief paragraph that compares and contrasts the ancient Indian and Chinese views of matter. How are they different? Similar? Which is closer to the modern view of matter? Explain.

Science Online

For more information, visit bookk.msscience.com/time

Reviewing Main Ideas

Section 1 Models of the Atom

1. Matter is made up of very small particles called atoms.

2. Atoms are made of smaller parts called protons, neutrons, and electrons.

3. Many models of atoms have been created as scientists try to discover and define the atom's internal structure. Today's model has a central nucleus with the protons and neutrons, and an electron cloud surrounding it that contains the electrons.

Section 2 The Simplest Matter

1. Elements are the basic building blocks of matter.

2. An element's atomic number tells how many protons its atoms contain, and its atomic mass tells the average atomic mass of its atoms.

3. Isotopes are two or more atoms of the same element that have different numbers of neutrons.

Section 3 Compounds and Mixtures

1. Compounds are substances that are produced when elements combine. Compounds contain specific proportions of the elements that make them up.

2. Mixtures are combinations of compounds and elements that have not formed new substances. Their proportions can change.

Visualizing Main Ideas

Copy and complete this concept map.

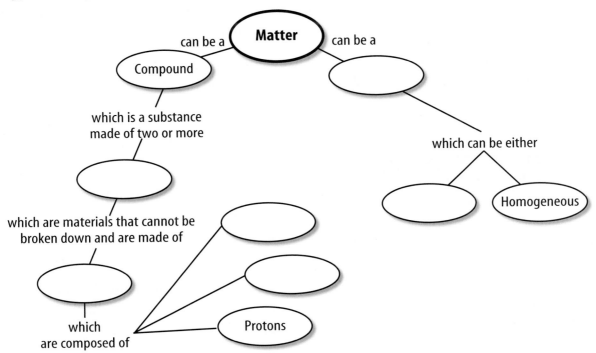

atomic mass p. 22	metal p. 22
atomic number p. 21	metalloid p. 23
compound p. 25	mixture p. 27
electron p. 11	neutron p. 15
electron cloud p. 17	nonmetal p. 23
element p. 9	proton p. 14
isotope p. 21	substance p. 25
mass number p. 21	

Fill in the blanks with the correct word.

1. The _____ is the particle in the nucleus of the atom that carries a positive charge and is counted to identify the atomic number.

2. The new substance formed when elements combine chemically is a(n) _____.

3. The _____ is equal to the number of protons in an atom.

4. The particles in the atom that account for most of the mass of the atom are protons and _____.

5. Elements that are shiny, malleable, ductile, good conductors of heat and electricity, and make up most of the periodic table are _____.

Checking Concepts

Choose the word or phrase that best answers the question.

6. What is a solution an example of?
   A) element
   B) heterogeneous mixture
   C) compound
   D) homogeneous mixture

7. The nucleus of one atom contains 12 protons and 12 neutrons, while the nucleus of another atom contains 12 protons and 16 neutrons. What are the atoms?
   A) chromium atoms
   B) two different elements
   C) two isotopes of an element
   D) negatively charged

8. What is a compound?
   A) a mixture of chemicals and elements
   B) a combination of two or more elements
   C) anything that has mass and occupies space
   D) the building block of matter

9. What does the atom consist of?
   A) electrons, protons, and alpha particles
   B) neutrons and protons
   C) electrons, protons, and neutrons
   D) elements, protons, and electrons

10. In an atom, where is an electron located?
    A) in the nucleus with the proton
    B) on the periodic table of the elements
    C) with the neutron
    D) in a cloudlike formation surrounding the nucleus

11. How is mass number defined?
    A) the negative charge in an atom
    B) the number of protons and neutrons in an atom
    C) the mass of the nucleus
    D) an atom's protons

12. What are two atoms that have the same number of protons called?
    A) metals   C) isotopes
    B) nonmetals   D) metalloids

13. Which is a heterogeneous mixture?
    A) air   C) a salad
    B) brass   D) apple juice

Use the illustration below to answer questions 14 and 15.

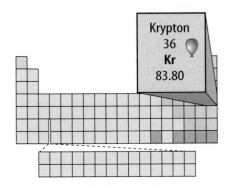

Krypton
36
Kr
83.80

14. According to the figure above, krypton has
 A) an atomic number of 84.
 B) an atomic number of 36.
 C) an atomic mass of 36.
 D) an atomic mass of 72.

15. From the figure, the element krypton is
 A) a solid. **C)** a mixture.
 B) a liquid. **D)** a gas.

Thinking Critically

16. **Analyze Information** A chemical formula is written to indicate the makeup of a compound. What is the ratio of sulfur atoms to oxygen atoms in SO_2?

17. **Determine** which element contains seven electrons and seven protons. What element is this atom?

18. **Describe** what happens to an element when it becomes part of a compound.

19. **Explain** how cobalt-60 and cobalt-59 can be the same element but have different mass numbers.

20. **Analyze Information** What did Rutherford's gold foil experiment tell scientists about atomic structure?

21. **Predict** Suppose Rutherford had bombarded aluminum foil with alpha particles instead of the gold foil he used in his experiment. What observations do you predict Rutherford would have made? Explain your prediction.

22. **Compare and Contrast** Aluminum is close to carbon on the periodic table. List the properties that make aluminum a metal and carbon a nonmetal.

23. **Draw Conclusions** You are shown a liquid that looks the same throughout. You're told that it contains more than one type of element and that the proportion of each varies throughout the liquid. Is this an element, a compound, or a mixture.

Use the illustration below to answer question 24.

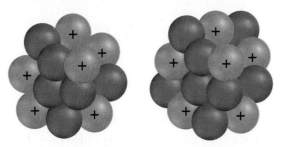

24. **Interpret Scientific Illustrations** Look at the two carbon atoms above. Explain whether or not the atoms are isotopes.

25. **Explain** how the atomic mass of krypton was determined.

Performance Activities

26. **Newspaper Article** Research the source, composition, and properties of asbestos. Why was it used in the past? Why is it a health hazard now? What is being done about it? Write a newspaper article to share your findings.

Applying Math

27. **Calculate** Krypton has six naturally occurring isotopes with atomic masses of 78, 80, 82, 83, 84, and 86. Make a table of the number of protons, electrons, and neutrons in each isotope.

Part 1 Multiple Choice

Record your answers on the answer sheet provided by your teacher or on a sheet of paper.

1. Which of the following has the smallest size?
  A. electron
  C. proton
  B. nucleus
  D. neutron

Use the illustration below to answer questions 2 and 3.

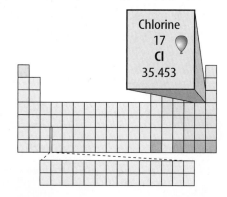

2. The periodic table block shown above lists properties of the element chlorine. What does the number balloon mean?
  A. gas
  B. liquid
  C. solid
  D. synthetic

3. According to the periodic table block, how many electrons does an uncharged atom of chlorine have?
  A. 17
  B. 18
  C. 35
  D. 36

Test-Taking Tip

Answer Each Question Never leave any constructed-response answer blank. Answer each question as best as you can. You can receive partial credit for partially correct answers.

4. Which of the following scientists envisioned the atom having a hard sphere that is the same throughout?
  A. Crookes
  C. Thomson
  B. Dalton
  D. Rutherford

Use the illustration below to answer questions 5 and 6.

| 1 Proton | 1 Proton | 1 Proton |
| 0 Neutrons | 1 Neutron | 2 Neutrons |

5. Which of the following correctly identifies the three atoms shown in the illustration above?
  A. hydrogen, lithium, sodium
  B. hydrogen, helium, lithium
  C. hydrogen, hydrogen, hydrogen
  D. hydrogen, helium, helium

6. What is the mass number for each of the atoms shown in the illustration?
  A. 0, 1, 2
  C. 1, 2, 2
  B. 1, 1, 1
  D. 1, 2, 3

7. Which of the following are found close to the right side of the periodic table?
  A. metals
  C. nonmetals
  B. lanthanides
  D. metalloids

8. Which of the following best describes a neutron?
  A. positive charge; about the same mass as an electron
  B. no charge; about the same mass as a proton
  C. negative charge; about the same mass as a proton
  D. no charge; about the same mass as an electron

Part 2 | Short Response/Grid In

Record your answers on the answer sheet provided by your teacher or on a sheet of paper.

9. Are electrons more likely to be close to the nucleus or far away from the nucleus? Why?

10. How many naturally occurring elements are listed on the periodic table?

11. Is the human body made of mostly metal, nonmetals, or metalloids?

12. A molecule of hydrogen peroxide is composed of two atoms of hydrogen and two atoms of oxygen. What is the formula for six molecules of hydrogen peroxide?

13. What is the present-day name for cathode rays?

Use the illustration below to answer questions 14 and 15.

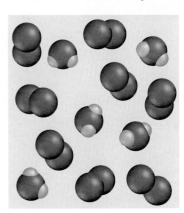

Enclosed sample of air

14. The illustration above shows atoms of an element and molecules of a compound that are combined without making a new compound. What term describes a combination such as this?

15. If the illustration showed only the element or only the compound, what term would describe it?

Part 3 | Open Ended

Record your answers on a sheet of paper.

16. Describe Dalton's ideas about the composition of matter, including the relationship between atoms and elements.

Use the illustration below to answer questions 17 and 18.

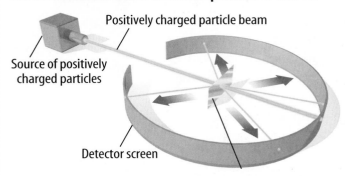

Positively charged particle beam

Source of positively charged particles

Detector screen

17. The illustration above shows Rutherford's gold foil experiment. Describe the setup shown. What result did Rutherford expect from his experiment?

18. What is the significance of the particles that reflected back from the gold foil? How did Rutherford explain his results?

19. Describe three possible methods for separating mixtures. Give an example for each method.

20. Describe the difference between a homogeneous and a heterogeneous mixture.

21. What are the rows and columns on the periodic table called? How are elements in the rows similar, and how are elements in the columns similar?

22. Describe how Thomson was able to show that cathode rays were streams of particles, not light.

23. Describe how the mass numbers, or atomic masses, listed on the periodic table for the elements are calculated.

States of Matter

The BIG Idea

The particles in solids, liquids, and gases are always in motion.

SECTION 1
Matter
Main Idea The state of matter depends on the motion of the particles and on the attractions between them.

SECTION 2
Changes of State
Main Idea When matter changes state, its thermal energy changes.

SECTION 3
Behavior of Fluids
Main Idea The particles in a fluid, a liquid, or a gas exert a force on everything they touch.

Ahhh!

A long, hot soak on a snowy day! This Asian monkey called a macaque is experiencing the effects of heat—the transfer of thermal energy from a warmer object to a colder object. In this chapter, you will learn about heat and the three common states of matter on Earth.

Science Journal Write why you think there is snow on the ground but the water is not frozen.

Start-Up Activities

Experiment with a Freezing Liquid

Have you ever thought about how and why you might be able to ice-skate on a pond in the winter but swim in the same pond in the summer? Many substances change form as temperature changes.

1. Make a table to record temperature and appearance. Obtain a test tube containing an unknown liquid from your teacher. Place the test tube in a rack.

2. Insert a thermometer into the liquid. **WARNING:** *Do not allow the thermometer to touch the bottom of the test tube.* Starting immediately, observe and record the substance's temperature and appearance every 30 s.

3. Continue making measurements and observations until you're told to stop.

4. **Think Critically** In your Science Journal, describe your investigation and observations. Did anything unusual happen while you were observing? If so, what?

Preview this chapter's content and activities at bookk.msscience.com

FOLDABLES™
Study Organizer

Changing States of Matter
Make the following Foldable to help you study the changes in water.

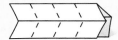

STEP 1 Fold a vertical sheet of paper from left to right two times. Unfold.

STEP 2 Fold the paper in half from top to bottom two times.

STEP 3 Unfold and draw lines along the folds.

STEP 4 Label the top row and first column as shown below.

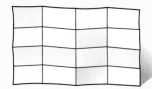

	Define States	+ Heat	– Heat
Liquid water			
Water as a gas			
Water as a solid (ice)			

Read and Write As you read the chapter, define the states of matter as listed on your Foldable in the *Define States* column. Write what happens when heat is added to or lost from the three states of matter.

Get Ready to Read

① Learn It! An important strategy to help you improve your reading is monitoring, or finding your reading strengths and weaknesses. As you read, monitor yourself to make sure the text makes sense. Discover different monitoring techniques you can use at different times, depending on the type of test and situation.

② Practice It! The paragraph below appears in Section 1. Read the passage and answer the questions that follow. Discuss your answers with other students to see how they monitor their reading.

> All matter is made up of tiny particles, such as atoms, molecules, or ions. Each particle attracts other particles. In other words, each particle pulls other particles toward itself. These particles also are constantly moving. The motion of the particles and the strength of attraction between the particles determine a material's state of matter.
>
> — *from page 40*

- What questions do you still have after reading?
- Do you understand all of the words in the passage?
- Did you have to stop reading often? Is the reading level appropriate for you?

③ Apply It! Identify one paragraph that is difficult to understand. Discuss it with a partner to improve your understanding.

Target Your Reading

Reading Tip

Monitor your reading by slowing down or speeding up depending on your understanding of the text.

Use this to focus on the main ideas as you read the chapter.

1 **Before you read** the chapter, respond to the statements below on your worksheet or on a numbered sheet of paper.

- Write an **A** if you **agree** with the statement.
- Write a **D** if you **disagree** with the statement.

2 **After you read** the chapter, look back to this page to see if you've changed your mind about any of the statements.

- If any of your answers changed, explain why.
- Change any false statements into true statements.
- Use your revised statements as a study guide.

Science Online

Print out a worksheet of this page at bookk.msscience.com

Before You Read A or D		Statement	After You Read A or D
	1	Particles in solids vibrate in place.	
	2	A water spider can walk on water because of uneven forces acting on the surface water molecules.	
	3	Particles in a gas are far apart with empty space between them.	
	4	A large glass of warm water has the same amount of thermal energy as a smaller glass of water at the same temperature.	
	5	Boiling and evaporation are two types of vaporization.	
	6	While a substance is boiling, its temperature increases.	
	7	Pressure is, in part, related to the area over which a force is distributed.	
	8	At sea level, the air exerts a pressure of about 101,000 N per square meter.	
	9	An object will float in a fluid that is denser than itself.	

Matter

What You'll Learn

- **Recognize** that matter is made of particles in constant motion.
- **Relate** the three states of matter to the arrangement of particles within them.

Why It's Important

Everything you can see, taste, and touch is matter.

Review Vocabulary

atom: a small particle that makes up most types of matter

New Vocabulary

- matter
- solid
- liquid
- viscosity
- surface tension
- gas

What is matter?

Take a look at the beautiful scene in **Figure 1.** What do you see? Perhaps you notice the water and ice. Maybe you are struck by the Sun in the background. All of these images show examples of matter. **Matter** is anything that takes up space and has mass. Matter doesn't have to be visible—even air is matter.

States of Matter All matter is made up of tiny particles, such as atoms, molecules, or ions. Each particle attracts other particles. In other words, each particle pulls other particles toward itself. These particles also are constantly moving. The motion of the particles and the strength of attraction between the particles determine a material's state of matter.

Reading Check *What determines a material's state of matter?*

There are three familiar states of matter—solid, liquid, and gas. A fourth state of matter known as plasma occurs at extremely high temperatures. Plasma is found in stars, lightning, and neon lights. Although plasma is common in the universe, it is not common on Earth. For that reason, this chapter will focus only on the three states of matter that are common on Earth.

Figure 1 Matter exists in all four states in this scene.
Identify *the solid, liquid, gas, and plasma in this photograph.*

Solids

What makes a substance a solid? Think about some familiar solids. Chairs, floors, rocks, and ice cubes are a few examples of matter in the solid state. What properties do all solids share? A **solid** is matter with a definite shape and volume. For example, when you pick up a rock from the ground and place it in a bucket, it doesn't change shape or size. A solid does not take the shape of a container in which it is placed. This is because the particles of a solid are packed closely together, as shown in **Figure 2.**

Particles in Motion The particles that make up all types of matter are in constant motion. Does this mean that the particles in a solid are moving too? Although you can't see them, a solid's particles are vibrating in place. The particles do not have enough energy to move out of their fixed positions.

Reading Check *What motion do solid particles have?*

Crystalline Solids In some solids, the particles are arranged in a repeating, three-dimensional pattern called a crystal. These solids are called crystalline solids. In **Figure 3** you can see the arrangement of particles in a crystal of sodium chloride, which is table salt. The particles in the crystal are arranged in the shape of a cube. Diamond, another crystalline solid, is made entirely of carbon atoms that form crystals that look more like pyramids. Sugar, sand, and snow are other crystalline solids.

Solid

Figure 2 The particles in a solid vibrate in place while maintaining a constant shape and volume.

Figure 3 The particles in a crystal of sodium chloride (NaCl) are arranged in an orderly pattern.

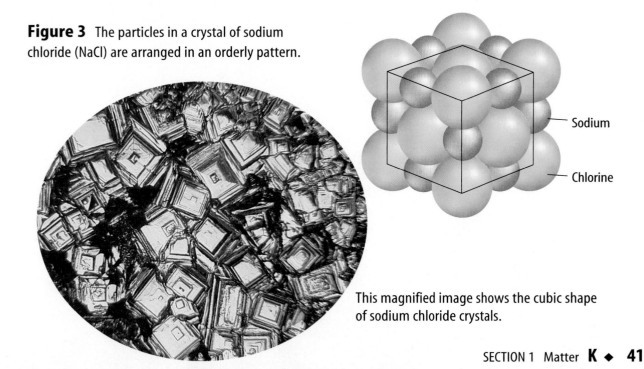

Sodium

Chlorine

This magnified image shows the cubic shape of sodium chloride crystals.

Amorphous Solids Some solids come together without forming crystal structures. These solids often consist of large particles that are not arranged in a repeating pattern. Instead, the particles are found in a random arrangement. These solids are called amorphous (uh MOR fuhs) solids. Rubber, plastic, and glass are examples of amorphous solids.

✓ Reading Check *How is a crystalline solid different from an amorphous solid?*

Liquids

From the orange juice you drink with breakfast to the water you use to brush your teeth at night, matter in the liquid state is familiar to you. How would you describe the characteristics of a liquid? Is it hard like a solid? Does it keep its shape? A **liquid** is matter that has a definite volume but no definite shape. When you pour a liquid from one container to another, the liquid takes the shape of the container. The volume of a liquid, however, is the same no matter what the shape of the container. If you pour 50 mL of juice from a carton into a pitcher, the pitcher will contain 50 mL of juice. If you then pour that same juice into a glass, its shape will change again but its volume will not.

Free to Move The reason that a liquid can have different shapes is because the particles in a liquid move more freely, as shown in **Figure 4,** than the particles in a solid. The particles in a liquid have enough energy to move out of their fixed positions but not enough energy to move far apart.

Figure 4 The particles in a liquid stay close together, although they are free to move past one another.

Liquid

Viscosity Do all liquids flow the way water flows? You know that honey flows more slowly than water and you've probably heard the phrase "slow as molasses." Some liquids flow more easily than others. A liquid's resistance to flow is known as the liquid's **viscosity.** Honey has a high viscosity. Water has a lower viscosity. The slower a liquid flows, the higher its viscosity is. The viscosity results from the strength of the attraction between the particles of the liquid. For many liquids, viscosity increases as the liquid becomes colder.

Surface Tension If you're careful, you can float a needle on the surface of water. This is because attractive forces cause the particles on the surface of a liquid to pull themselves together and resist being pushed apart. You can see in **Figure 5** that particles beneath the surface of a liquid are pulled in all directions. Particles at the surface of a liquid are pulled toward the center of the liquid and sideways along the surface. No liquid particles are located above to pull on them. The uneven forces acting on the particles on the surface of a liquid are called **surface tension.** Surface tension causes the liquid to act as if a thin film were stretched across its surface. As a result you can float a needle on the surface of water. For the same reason, the water spider can move around on the surface of a pond or lake. When a liquid is present in small amounts, surface tension causes the liquid to form small droplets.

Science nline

Topic: Plasma
Visit bookk.msscience.com for Web links to information about the states of matter.

Activity List four ways that plasma differs from the other three states of matter

Figure 5 Surface tension exists because the particles at the surface experience different forces than those at the center of the liquid.

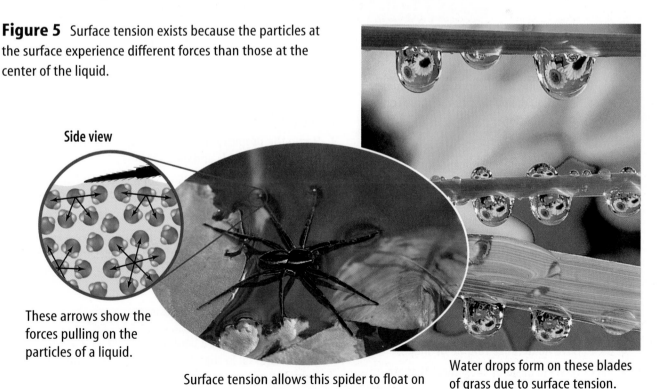

Side view

These arrows show the forces pulling on the particles of a liquid.

Surface tension allows this spider to float on water as if the water had a thin film.

Water drops form on these blades of grass due to surface tension.

Gases

Unlike solids and liquids, most gases are invisible. The air you breathe is a mixture of gases. The gas in the air bags in **Figure 6** and the helium in some balloons are examples of gases. **Gas** is matter that does not have a definite shape or volume. The particles in gas are much farther apart than those in a liquid or solid. Gas particles move at high speeds in all directions. They will spread out evenly, as far apart as possible. If you poured a small volume of a liquid into a container, the liquid would stay in the bottom of the container. However, if you poured the same volume of a gas into a container, the gas would fill the container completely. A gas can expand or be compressed. Decreasing the volume of the container squeezes the gas particles closer together.

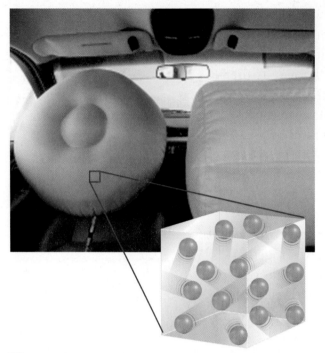

Figure 6 The particles in gas move at high speeds in all directions. The gas inside these air bags spreads out to fill the entire volume of the bag.

Vapor Matter that exists in the gas state but is generally a liquid or solid at room temperature is called vapor. Water, for example, is a liquid at room temperature. Thus, water vapor is the term for the gas state of water.

section 1 review

Summary

What is matter?

● Matter is anything that takes up space and has mass. Solid, liquid, and gas are the three common states of matter.

Solids

● Solids have a definite volume and shape.

● Solids with particles arranged in order are called crystalline solids. The particles in amorphous solids are not in any order.

Liquids

● Liquids have definite volume but no defined shape.

● Viscosity is a measure of how easily liquids flow.

Gases

● Gases have no definite volume or shape.

● Vapor refers to gaseous substances that are normally liquids or solids at room temperature.

Self Check

1. **Define** the two properties of matter that determine its state.

2. **Describe** the movement of particles within solids, liquids, and gases.

3. **Name** the property that liquids and solids share. What property do liquids and gases share?

4. **Infer** A scientist places 25 mL of a yellow substance into a 50-mL container. The substance quickly fills the entire container. Is it a solid, liquid, or gas?

5. **Think Critically** The particles in liquid A have a stronger attraction to each other than the particles in liquid B. If both liquids are at the same temperature, which liquid has a higher viscosity? Explain.

Applying Skills

6. **Concept Map** Draw a Venn diagram in your Science Journal and fill in the characteristics of the states of matter.

Changes of State

Thermal Energy and Heat

Shards of ice fly from the sculptor's chisel. As the crowd looks on, a swan slowly emerges from a massive block of ice. As the day wears on, however, drops of water begin to fall from the sculpture. Drip by drip, the sculpture is transformed into a puddle of liquid water. What makes matter change from one state to another? To answer this question, you need to think about the particles that make up matter.

Energy Simply stated, energy is the ability to do work or cause change. The energy of motion is called kinetic energy. Particles within matter are in constant motion. The amount of motion of these particles depends on the kinetic energy they possess. Particles with more kinetic energy move faster and farther apart. Particles with less energy move more slowly and stay closer together.

The total kinetic and potential energy of all the particles in a sample of matter is called **thermal energy.** Thermal energy, an extensive property, depends on the number of particles in a substance as well as the amount of energy each particle has. If either the number of particles or the amount of energy in each particle changes, the thermal energy of the sample changes. With identically sized samples, the warmer substance has the greater thermal energy. In **Figure 7,** the particles of hot water from the hot spring have more thermal energy than the particles of snow on the surrounding ground.

as you read

What You'll Learn
- **Define and compare** thermal energy and temperature.
- **Relate** changes in thermal energy to changes of state.
- **Explore** energy and temperature changes on a graph.

Why It's Important

Matter changes state as it heats up or cools down.

Review Vocabulary
energy: the ability to do work or cause change

New Vocabulary
- thermal energy
- temperature
- heat
- melting
- freezing
- vaporization
- condensation

Figure 7 These girls are enjoying the water from the hot spring. **Infer** *why the girls appear to be comfortable in the hot spring while there is snow on the ground.*

Figure 8 The particles in hot tea move faster than those in iced tea. The temperature of hot tea is higher than the temperature of iced tea.
Identify *which tea has the higher kinetic energy.*

Types of Energy Thermal energy is one of several different forms of energy. Other forms include the chemical energy in chemical compounds, the electrical energy used in appliances, the electromagnetic energy of light, and the nuclear energy stored in the nucleus of an atom. Make a list of examples of energy that you are familiar with.

Temperature Not all of the particles in a sample of matter have the same amount of energy. Some have more energy than others. The average kinetic energy of the individual particles is the **temperature,** an intensive property, of the substance. You can find an average by adding up a group of numbers and dividing the total by the number of items in the group. For example, the average of the numbers 2, 4, 8, and 10 is $(2 + 4 + 8 + 10) \div 4 = 6$. Temperature is different from thermal energy because thermal energy is a total and temperature is an average.

You know that the iced tea is colder than the hot tea, as shown in **Figure 8.** Stated differently, the temperature of iced tea is lower than the temperature of hot tea. You also could say that the average kinetic energy of the particles in the iced tea is less than the average kinetic energy of the particles in the hot tea.

Heat When a warm object is brought near a cooler object, thermal energy will be transferred from the warmer object to the cooler one. The movement of thermal energy from a substance at a higher temperature to one at a lower temperature is called **heat.** When a substance is heated, it gains thermal energy. Therefore, its particles move faster and its temperature rises. When a substance is cooled, it loses thermal energy, which causes its particles to move more slowly and its temperature to drop.

✔ **Reading Check** *How is heat related to temperature?*

Specific Heat

As you study more science, you will discover that water has many unique properties. One of those is the amount of heat required to increase the temperature of water as compared to most other substances. The specific heat of a substance is the amount of heat required to raise the temperature of 1 g of a substance 1°C.

Substances that have a low specific heat, such as most metals and the sand in **Figure 9,** heat up and cool down quickly because they require only small amounts of heat to cause their temperatures to rise. A substance with a high specific heat, such as the water in **Figure 9,** heats up and cools down slowly because a much larger quantity of heat is required to cause its temperature to rise or fall by the same amount.

Changes Between the Solid and Liquid States

Matter can change from one state to another when thermal energy is absorbed or released. This change is known as change of state. The graph in **Figure 11** shows the changes in temperature as thermal energy is gradually added to a container of ice.

Melting As the ice in **Figure 11** is heated, it absorbs thermal energy and its temperature rises. At some point, the temperature stops rising and the ice begins to change into liquid water. The change from the solid state to the liquid state is called **melting.** The temperature at which a substance changes from a solid to a liquid is called the melting point. The melting point of water is 0°C.

Amorphous solids, such as rubber and glass, don't melt in the same way as crystalline solids. Because they don't have crystal structures to break down, these solids get softer and softer as they are heated, as you can see in **Figure 10.**

Figure 9 The specific heat of water is greater than that of sand. The energy provided by the Sun raises the temperature of the sand much faster than the water.

Figure 10 Rather than melting into a liquid, glass gradually softens. Glass blowers use this characteristic to shape glass into beautiful vases while it is hot.

NATIONAL GEOGRAPHIC VISUALIZING STATES OF MATTER

Figure 11

Like most substances, water can exist in three distinct states—solid, liquid, or gas. At certain temperatures, water changes from one state to another. This diagram shows what changes occur as water is heated or cooled.

MELTING When ice melts, its temperature remains constant until all the ice turns to water. Continued heating of liquid water causes the molecules to vibrate even faster, steadily raising the temperature.

FREEZING When liquid water freezes, it releases thermal energy and turns into the solid state, ice.

VAPORIZATION When water reaches its boiling point of 100°C, water molecules are moving so fast that they break free of the attractions that hold them together in the liquid state. The result is vaporization—the liquid becomes a gas. The temperature of boiling water remains constant until all of the liquid turns to steam.

Gas

Vaporization

Condensation

Liquid

CONDENSATION When steam is cooled, it releases thermal energy and turns into its liquid state. This process is called condensation.

100°C

0°C

Temperature

Melting

Freezing

Solid

Thermal energy

Solid state: ice

Liquid state: water

Gaseous state: steam

Freezing The process of melting a crystalline solid can be reversed if the liquid is cooled. The change from the liquid state to the solid state is called **freezing.** As the liquid cools, it loses thermal energy. As a result, its particles slow down and come closer together. Attractive forces begin to trap particles, and the crystals of a solid begin to form. As you can see in **Figure 11,** freezing and melting are opposite processes.

The temperature at which a substance changes from the liquid state to the solid state is called the freezing point. The freezing point of the liquid state of a substance is the same temperature as the melting point of the solid state. For example, solid water melts at 0°C and liquid water freezes at 0°C.

During freezing, the temperature of a substance remains constant while the particles in the liquid form a crystalline solid. Because particles in a liquid have more energy than particles in a solid, energy is released during freezing. This energy is released into the surroundings. After all of the liquid has become a solid, the temperature begins to decrease again.

Science nline

Topic: Freezing Point Study
Visit bookk.msscience.com for Web links to information about freezing.

Activity Make a list of several substances and the temperatures at which they freeze. Find out how the freezing point affects how the substance is used.

Applying Science

How can ice save oranges?

During the spring, Florida citrus farmers carefully watch the fruit when temperatures drop close to freezing. When the temperatures fall below 0°C, the liquid in the cells of oranges can freeze and expand. This causes the cells to break, making the oranges mushy and the crop useless for sale. To prevent this, farmers spray the oranges with water just before the temperature reaches 0°C. How does spraying oranges with water protect them?

Identifying the Problem
Using the diagram in **Figure 11,** consider what is happening to the water at 0°C. Two things occur. What are they?

Solving the Problem
1. What change of state and what energy changes occur when water freezes?
2. How does the formation of ice on the orange help the orange?

Mini LAB

Observing Vaporization

Procedure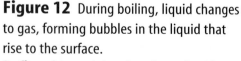

1. Use a **dropper** to place one drop of **rubbing alcohol** on the back of your hand.
2. Describe how your hand feels during the next 2 min.
3. Wash your hands.

Analysis

1. What changes in the appearance of the rubbing alcohol did you notice?
2. What sensation did you feel during the 2 min? How can you explain this sensation?
3. Infer how sweating cools the body.

Changes Between the Liquid and Gas States

After an early morning rain, you and your friends enjoy stomping through the puddles left behind. But later that afternoon when you head out to run through the puddles once more, the puddles are gone. The liquid water in the puddles changed into a gas. Matter changes between the liquid and gas states through vaporization and condensation.

Vaporization As liquid water is heated, its temperature rises until it reaches 100°C. At this point, liquid water changes into water vapor. The change from a liquid to a gas is known as **vaporization** (vay puh ruh ZAY shun). You can see in **Figure 11** that the temperature of the substance does not change during vaporization. However, the substance absorbs thermal energy. The additional energy causes the particles to move faster until they have enough energy to escape the liquid as gas particles.

Two forms of vaporization exist. Vaporization that takes place below the surface of a liquid is called boiling. When a liquid boils, bubbles form within the liquid and rise to the surface, as shown in **Figure 12.** The temperature at which a liquid boils is called the boiling point. The boiling point of water is 100°C.

Vaporization that takes place at the surface of a liquid is called evaporation. Evaporation, which occurs at temperatures below the boiling point, explains how puddles dry up. Imagine that you could watch individual water molecules in a puddle. You would notice that the molecules move at different speeds. Although the temperature of the water is constant, remember that temperature is a measure of the average kinetic energy of the molecules. Some of the fastest-moving molecules overcome the attractive forces of other molecules and escape from the surface of the water.

Figure 12 During boiling, liquid changes to gas, forming bubbles in the liquid that rise to the surface.
Define *the word that describes a liquid changing to the gas.*

Location of Molecules It takes more than speed for water molecules to escape the liquid state. During evaporation, these faster molecules also must be near the surface, heading in the right direction, and they must avoid hitting other water molecules as they leave. With the faster particles evaporating from the surface of a liquid, the particles that remain are the slower, cooler ones. Evaporation cools the liquid and anything near the liquid. You experience this cooling effect when perspiration evaporates from your skin.

Condensation Pour a nice, cold glass of lemonade and place it on the table for a half hour on a warm day. When you come back to take a drink, the outside of the glass will be covered by drops of water, as shown in **Figure 13.** What happened? As a gas cools, its particles slow down. When particles move slowly enough for their attractions to bring them together, droplets of liquid form. This process, which is the opposite of vaporization, is called **condensation.** As a gas condenses to a liquid, it releases the thermal energy it absorbed to become a gas. During this process, the temperature of the substance does not change. The decrease in energy changes the arrangement of particles. After the change of state is complete, the temperature continues to drop, as you saw in **Figure 11.**

 *What energy change occurs during condensation?*

Condensation formed the droplets of water on the outside of your glass of lemonade. In the same way, water vapor in the atmosphere condenses to form the liquid water droplets in clouds. When the droplets become large enough, they can fall to the ground as rain.

Science Online

Topic: Condensation
Visit bookk.msscience.com for Web links to information about how condensation is involved in weather.

Activity Find out how condensation is affected by the temperature as well as the amount of water in the air.

Changes Between the Solid and Gas States

Some substances can change from the solid state to the gas state without ever becoming a liquid. During this process, known as sublimation, the surface particles of the solid gain enough energy to become a gas. One example of a substance that undergoes sublimation is dry ice. Dry ice is the solid form of carbon dioxide. It often is used to keep materials cold and dry. At room temperature and pressure, carbon dioxide does not exist as a liquid. Therefore, as dry ice absorbs thermal energy from the objects around it, it changes directly into a gas. When dry ice becomes a gas, it absorbs thermal energy from water vapor in the air. As a result, the water vapor cools and condenses into liquid water droplets, forming the fog you see in **Figure 14.**

Figure 14 The solid carbon dioxide (dry ice) at the bottom of this beaker of water is changing directly into gaseous carbon dioxide. This process is called sublimation.

section 2 review

Summary

Thermal Energy and Heat

- Thermal energy depends on the amount of the substance and the kinetic energy of particles in the substance.
- Heat is the movement of thermal energy from a warmer substance to a cooler one.

Specific Heat

- Specific heat is a measure of the amount of energy required to raise 1 g of a substance 1°C.

Changes Between Solid and Liquid States

- During all changes of state, the temperature of a substance stays the same.

Changes Between Liquid and Gas States

- Vaporization is the change from the liquid state to a gaseous state.
- Condensation is the change from the gaseous state to the liquid state.

Changes Between Solid and Gas States

- Sublimation is the process of a substance going from the solid state to the gas state without ever being in the liquid state.

Self Check

1. **Describe** how thermal energy and temperature are similar. How are they different?

2. **Explain** how a change in thermal energy causes matter to change from one state to another. Give two examples.

3. **List** the three changes of state during which energy is absorbed.

4. **Describe** the two types of vaporization.

5. **Think Critically** How can the temperature of a substance remain the same even if the substance is absorbing thermal energy?

6. **Write** a paragraph in your Science Journal that explains why you can step out of the shower into a warm bathroom and begin to shiver.

Applying Math

7. **Make and Use Graphs** Use the data you collected in the Launch Lab to plot a temperature-time graph. At what temperature does the graph level off? What was the liquid doing during this time period?

8. **Use Numbers** If sample A requires 10 calories to raise the temperature of a 1-g sample 1°C, how many calories does it take to raise a 5-g sample 10°C?

The Water Cycle

Water is all around us and you've used water in all three of its common states. This lab will give you the opportunity to observe the three states of matter and to discover for yourself if ice really melts at 0°C and if water boils at 100°C.

◉ Real-World Question

How does the temperature of water change as it is heated from a solid to a gas?

Goals

■ **Measure** the temperature of water as it heats.

■ **Observe** what happens as the water changes from one state to another.

■ **Graph** the temperature and time data.

Materials

hot plate
ice cubes (100 mL)
Celsius thermometer
*electronic
    temperature probe
wall clock

*watch with
    second hand
stirring rod
250-mL beaker
*Alternate materials

Safety Precautions

◉ Procedure

1. Make a data table similar to the table shown.

2. Put 150 mL of water and 100 mL of ice into the beaker and place the beaker on the hot plate. Do not touch the hot plate.

3. Put the thermometer into the ice/water mixture. Do not stir with the thermometer or allow it to rest on the bottom of the beaker. After 30 s, read and record the temperature in your data table.

Characteristics of Water Sample		
Time (min)	Temperature (°C)	Physical State
	Do not write in this book.	

4. Plug in the hot plate and turn the temperature knob to the medium setting.

5. Every 30 s, read and record the temperature and physical state of the water until it begins to boil. Use the stirring rod to stir the contents of the beaker before making each temperature measurement. Stop recording. Allow the water to cool.

◉ Analyze Your Data

Use your data to make a graph plotting time on the x-axis and temperature on the y-axis. Draw a smooth curve through the data points.

◉ Conclude and Apply

1. **Describe** how the temperature of the ice/water mixture changed as you heated the beaker.

2. **Describe** the shape of the graph during any changes of state.

Communicating Your Data

Add labels to your graph. Use the detailed graph to explain to your class how water changes state. **For more help, refer to the Science Skill Handbook.**

Behavior of Fluids

What You'll Learn

- **Explain** why some things float but others sink.
- **Describe** how pressure is transmitted through fluids.

Why It's Important

Pressure enables you to squeeze toothpaste from a tube, and buoyant force helps you float in water.

⊙ Review Vocabulary

force: a push or pull

New Vocabulary

- pressure
- buoyant force
- Archimedes' principle
- density
- Pascal's principle

Pressure

It's a beautiful summer day when you and your friends go outside to play volleyball, much like the kids in **Figure 15.** There's only one problem—the ball is flat. You pump air into the ball until it is firm. The firmness of the ball is the result of the motion of the air particles in the ball. As the air particles in the ball move, they collide with one another and with the inside walls of the ball. As each particle collides with the inside walls, it exerts a force, pushing the surface of the ball outward. A force is a push or a pull. The forces of all the individual particles add together to make up the pressure of the air.

Pressure is equal to the force exerted on a surface divided by the total area over which the force is exerted.

$$\text{pressure} = \frac{\text{force}}{\text{area}}$$

When force is measured in newtons (N) and area is measured in square meters (m²), pressure is measured in newtons per square meter (N/m²). This unit of pressure is called a pascal (Pa). A more useful unit when discussing atmospheric pressure is the kilopascal (kPa), which is 1,000 pascals.

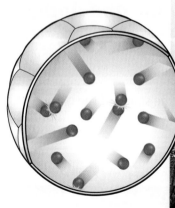

Figure 15 Without the pressure of air inside this volleyball, the ball would be flat.

Figure 16 The force of the dancer's weight on pointed toes results in a higher pressure than the same force on flat feet. **Explain** *why the pressure is higher.*

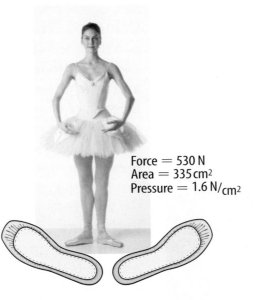

Force = 530 N
Area = 335 cm²
Pressure = 1.6 N/cm²

Force = 530 N
Area = 37 cm²
Pressure = 14 N/cm²

Force and Area You can see from the equation on the opposite page that pressure depends on the quantity of force exerted and the area over which the force is exerted. As the force increases over a given area, pressure increases. If the force decreases, the pressure will decrease. However, if the area changes, the same amount of force can result in different pressure. **Figure 16** shows that if the force of the ballerina's weight is exerted over a smaller area, the pressure increases. If that same force is exerted over a larger area, the pressure will decrease.

> ✔ **Reading Check** *What variables does pressure depend on?*

Atmospheric Pressure You can't see it and you usually can't feel it, but the air around you presses on you with tremendous force. The pressure of air also is known as atmospheric pressure because air makes up the atmosphere around Earth. Atmospheric pressure is 101.3 kPa at sea level. This means that air exerts a force of about 101,000 N on every square meter it touches. This is approximately equal to the weight of a large truck.

It might be difficult to think of air as having pressure when you don't notice it. However, you often take advantage of air pressure without even realizing it. Air pressure, for example, enables you to drink from a straw. When you first suck on a straw, you remove the air from it. As you can see in **Figure 17,** air pressure pushes down on the liquid in your glass then forces liquid up into the straw. If you tried to drink through a straw inserted into a sealed, airtight container, you would not have any success because the air would not be able to push down on the surface of the drink.

Figure 17 The downward pressure of air pushes the juice up into the straw.

Air pressure

Balanced Pressure

If air is so forceful, why don't you feel it? The reason is that the pressure exerted outward by the fluids in your body balances the pressure exerted by the atmosphere on the surface of your body. Look at **Figure 18.** The atmosphere exerts a pressure on all surfaces of the dancer's body. She is not crushed by this pressure because the fluids in her body exert a pressure that balances atmospheric pressure.

Variations in Atmospheric Pressure

Atmospheric pressure changes with altitude. Altitude is the height above sea level. As altitude increases atmospheric pressure decreases. This is because fewer air particles are found in a given volume. Fewer particles have fewer collisions, and therefore exert less pressure. This idea was tested in the seventeenth century by a French physician named Blaise Pascal. He designed an experiment in which he filled a balloon only partially with air. He then had the balloon carried to the top of a mountain. **Figure 19** shows that as Pascal predicted, the balloon expanded while being carried up the mountain. Although the amount of air inside the balloon stayed the same, the air pressure pushing in on it from the outside decreased. Consequently, the particles of air inside the balloon were able to spread out further.

Figure 18 Atmospheric pressure exerts a force on all surfaces of this dancer's body.
Explain *why she can't feel this pressure.*

Figure 19 Notice how the balloon expands as it is carried up the mountain. The reason is that atmospheric pressure decreases with altitude. With less pressure pushing in on the balloon, the gas particles within the balloon are free to expand.

Air Travel If you travel to higher altitudes, perhaps flying in an airplane or driving up a mountain, you might feel a popping sensation in your ears. As the air pressure drops, the air pressure in your ears becomes greater than the air pressure outside your body. The release of some of the air trapped inside your ears is heard as a pop. Airplanes are pressurized so that the air pressure within the cabin does not change dramatically throughout the course of a flight.

Changes in Gas Pressure

In the same way that atmospheric pressure can vary as conditions change, the pressure of gases in confined containers also can change. The pressure of a gas in a closed container changes with volume and temperature.

Pressure and Volume If you squeeze a portion of a filled balloon, the remaining portion of the balloon becomes more firm. By squeezing it, you decrease the volume of the balloon, forcing the same number of gas particles into a smaller space. As a result, the particles collide with the walls more often, thereby producing greater pressure. This is true as long as the temperature of the gas remains the same. You can see the change in the motion of the particles in **Figure 20.** What will happen if the volume of a gas increases? If you make a container larger without changing its temperature, the gas particles will collide less often and thereby produce a lower pressure.

Mini LAB

Predicting a Waterfall

Procedure

1. Fill a **plastic cup** to the brim with **water.**
2. Cover the top of the cup with an **index card.**
3. Predict what will happen if you turn the cup upside down.
4. While holding the index card in place, turn the cup upside down over a sink. Then let go of the card.

Analysis

1. What happened to the water when you turned the cup?
2. How can you explain your observation in terms of the concept of fluid pressure?

Try at Home

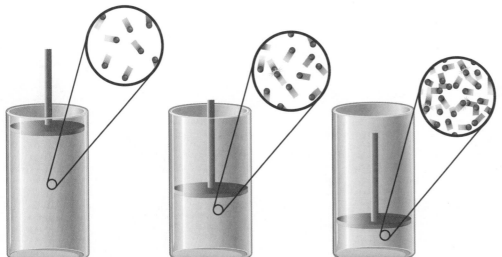

Figure 20 As volume decreases, pressure increases.

As the piston is moved down, the gas particles have less space and collide more often. The pressure increases.

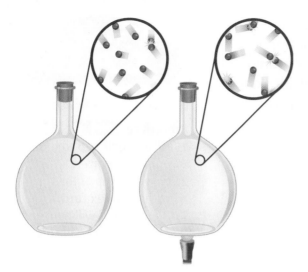

Pressure and Temperature When the volume of a confined gas remains the same, the pressure can change as the temperature of the gas changes. You have learned that temperature rises as the kinetic energy of the particles in a substance increases. The greater the kinetic energy is, the faster the particles move. The faster the speed of the particles is, the more they collide and the greater the pressure is. If the temperature of a confined gas increases, the pressure of the gas will increase, as shown in **Figure 21.**

Reading Check *Why would a sealed container of air be crushed after being frozen?*

Figure 21 Even though the volume of this container does not change, the pressure increases as the substance is heated.
Describe *what will happen if the substance is heated too much.*

Float or Sink

You may have noticed that you feel lighter in water than you do when you climb out of it. While you are under water, you experience water pressure pushing on you in all directions. Just as air pressure increases as you walk down a mountain, water pressure increases as you swim deeper in water. Water pressure increases with depth. As a result, the pressure pushing up on the bottom of an object is greater than the pressure pushing down on it because the bottom of the object is deeper than the top.

The difference in pressure results in an upward force on an object immersed in a fluid, as shown in **Figure 22.** This force is known as the **buoyant force.** If the buoyant force is equal to the weight of an object, the object will float. If the buoyant force is less than the weight of an object, the object will sink.

Figure 22 The pressure pushing up on an immersed object is greater than the pressure pushing down on it. This difference results in the buoyant force.

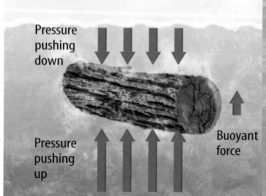

Weight is a force in the downward direction. The buoyant force is in the upward direction. An object will float if the upward force is equal to the downward force.

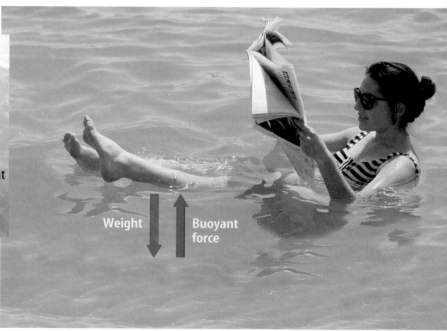

Archimedes' Principle What determines the buoyant force? According to **Archimedes'** (ar kuh MEE deez) **principle,** the buoyant force on an object is equal to the weight of the fluid displaced by the object. In other words, if you place an object in a beaker that already is filled to the brim with water, some water will spill out of the beaker, as in **Figure 23.** If you weigh the spilled water, you will find the buoyant force on the object.

Density Understanding density can help you predict whether an object will float or sink. **Density** is mass divided by volume.

$$\text{density} = \frac{\text{mass}}{\text{volume}}$$

An object will float in a fluid that is more dense than itself and sink in a fluid that is less dense than itself. If an object has the same density, the object will neither sink nor float but instead stay at the same level in the fluid.

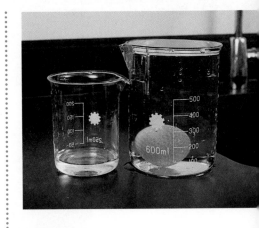

Figure 23 When the golf ball was dropped in the large beaker, it displaced some of the water, which was collected and placed into the smaller beaker. **Communicate** *what you know about the weight and the volume of the displaced water.*

Applying Math Find an Unknown

CALCULATING DENSITY You are given a sample of a solid that has a mass of 10.0 g and a volume of 4.60 cm³. Will it float in liquid water, which has a density of 1.00 g/cm³?

Solution

1 *This is what you know:*

- mass = 10.0 g
- volume = 4.60 cm³
- density of water = 1.00 g/cm³

2 *This is what you need to find:* the density of the sample

3 *This is the procedure you need to use:*

- density = mass/volume
- density = 10.0 g/4.60 cm³ = 2.17 g/cm³
- The density of the sample is greater than the density of water. The sample will sink.

4 *Check your answer:*

- Find the mass of your sample by multiplying the density and the volume.

Practice Problems

1. A 7.40-cm³ sample of mercury has a mass of 102 g. Will it float in water?

2. A 5.0-cm³ sample of aluminum has a mass of 13.5 g. Will it float in water?

 For more practice, visit bookk.msscience.com/ math_practice

Pascal's Principle

What happens if you squeeze a plastic container filled with water? If the container is closed, the water has nowhere to go. As a result, the pressure in the water increases by the same amount everywhere in the container—not just where you squeeze or near the top of the container. When a force is applied to a confined fluid, an increase in pressure is transmitted equally to all parts of the fluid. This relationship is known as **Pascal's principle.**

Hydraulic Systems You witness Pascal's principle when a car is lifted up to have its oil changed or if you are in a dentist's chair as it is raised or lowered, as shown in **Figure 24.** These devices, known as hydraulic (hi DRAW lihk) systems, use Pascal's principle to increase force. Look at the tube in **Figure 25.** The force applied to the piston on the left increases the pressure within the fluid. That increase in pressure is transmitted to the piston on the right. Recall that pressure is equal to force divided by area. You can solve for force by multiplying pressure by area.

$$\text{pressure} = \frac{\text{force}}{\text{area}} \quad \text{or} \quad \text{force} = \text{pressure} \times \text{area}$$

If the two pistons on the tube have the same area, the force will be the same on both pistons. If, however, the piston on the right has a greater surface area than the piston on the left, the resulting force will be greater. The same pressure multiplied by a larger area equals a greater force. Hydraulic systems enable people to lift heavy objects using relatively small forces.

Figure 24 A hydraulic lift utilizes Pascal's principle to help lift this car and this dentist's chair.

Figure 25 By increasing the area of the piston on the right side of the tube, you can increase the force exerted on the piston. In this way a small force pushing down on the left piston can result in a large force pushing up on the right piston. The force can be great enough to lift a car.

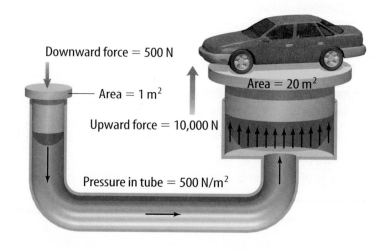

Downward force = 500 N

Area = 1 m²

Area = 20 m²

Upward force = 10,000 N

Pressure in tube = 500 N/m²

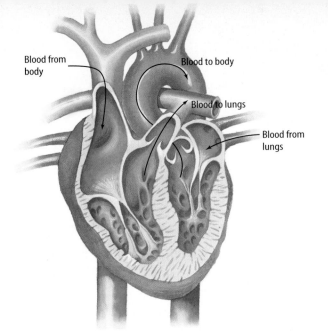

Blood from body

Blood to body

Blood to lungs

Blood from lungs

Figure 26 The heart is responsible for moving blood throughout the body. Two force pumps work together to move blood to and from the lungs and to the rest of the body.

Force Pumps If an otherwise closed container has a hole in it, any fluid in the container will be pushed out the opening when you squeeze it. This arrangement, known as a force pump, makes it possible for you to squeeze toothpaste out of a tube or mustard from a plastic container.

Your heart has two force pumps. One pump pushes blood to the lungs, where it picks up oxygen. The other force pump pushes the oxygen-rich blood to the rest of your body. These pumps are shown in **Figure 26.**

Topic: Blood Pressure
Visit bookk.msscience.com for Web links to information about blood pressure. Find out what the term means, how it changes throughout the human body, and why it is unhealthy to have high blood pressure.

Activity Write a paragraph in your Science Journal that explains why high blood pressure is dangerous.

section 3 review

Summary

Pressure
- Pressure depends on force and area.
- The air around you exerts a pressure.
- The pressure inside your body matches the pressure exerted by air.

Changes in Gas Pressure
- The pressure exerted by a gas depends on its volume and its temperature.

Float or Sink
- Whether an object floats or sinks depends on its density relative to the density of the fluid it's in.

Pascal's Principle
- This principle relates pressure and area to force.

Self Check

1. **Describe** what happens to pressure as the force exerted on a given area increases.
2. **Describe** how atmospheric pressure changes as altitude increases.
3. **State** Pascal's principle in your own words.
4. **Infer** An object floats in a fluid. What can you say about the buoyant force on the object?
5. **Think Critically** All the air is removed from a sealed metal can. After the air has been removed, the can looks as if it were crushed. Why?

Applying Math

6. **Simple Equations** What pressure is created when 5.0 N of force are applied to an area of 2.0 m^2? How does the pressure change if the force is increased to 10.0 N? What about if instead the area is decreased to 1.0 m^2?

Design Your Own

Design Your ⚓wn Ship

Goals

- **Design** an experiment that uses Archimedes' principle to determine the size of ship needed to carry a given amount of cargo in such a way that the top of the ship is even with the surface of the water.

Possible Materials

balance
small plastic cups (2)
graduated cylinder
metric ruler
scissors
marbles (cupful)
sink
*basin, pan, or bucket
*Alternate materials

Safety Precautions

▶ Real-World Question

It is amazing to watch ships that are taller than buildings float easily on water. Passengers and cargo are carried on these ships in addition to the tremendous weight of the ship itself. How can you determine the size of a ship needed to keep a certain mass of cargo afloat?

▶ Form a Hypothesis

Think about Archimedes' principle and how it relates to buoyant force. Form a hypothesis to explain how the volume of water displaced by a ship relates to the mass of cargo the ship can carry.

Cargo ship

▶ Test Your Hypothesis

Make a Plan

1. Obtain a set of marbles or other items from your teacher. This is the cargo that your ship must carry. Think about the type of ship

you will design. Consider the types of materials you will use. Decide how your group is going to test your hypothesis.

2. **List** the steps you need to follow to test your hypothesis. Include in your plan how you will measure the mass of your ship and cargo, calculate the volume of water your ship must displace in order to float with its cargo, and measure the volume and mass of the displaced water. Also, explain how you will design your ship so that it will float with the top of the ship even with the surface of the water. Make the ship.

3. **Prepare** a data table in your Science Journal to use as your group collects data. Think about what data you need to collect.

Follow Your Plan

1. Make sure your teacher approves your plan before you start.

2. Perform your experiment as planned. Be sure to follow all proper safety procedures. In particular, clean up any spilled water immediately.

3. Record your observations carefully and complete the data table in your Science Journal.

◉ *Analyze Your Data*

1. **Write** your calculations showing how you determined the volume of displaced water needed to make your ship and cargo float.

2. Did your ship float at the water's surface, sink, or float above the water's surface? Draw a diagram of your ship in the water.

3. **Explain** how your experimental results agreed or failed to agree with your hypothesis.

◉ *Conclude and Apply*

1. If your ship sank, how would you change your experiment or calculations to correct the problem? What changes would you make if your ship floated too high in the water?

2. What does the density of a ship's cargo have to do with the volume of cargo the ship can carry? What about the density of the water?

*C*ommunicating
Your Data

Compare your results with other students' data. Prepare a combined data table or summary showing how the calculations affect the success of the ship. **For more help, refer to the** Science Skill Handbook.

The Incredible Stretching Goo

A serious search turns up a toy

During World War II, when natural resources were scarce and needed for the war effort, the U.S. government asked an engineer to come up with an inexpensive alternative to synthetic rubber. While researching the problem and looking for solutions, the engineer dropped boric acid into silicone oil. The result of these two substances mixing together was—a goo!

Because of its molecular structure, the goo could bounce and stretch in all directions. The engineer also discovered the goo could break into pieces. When strong pressure is applied to the substance, it reacts like a solid and breaks apart. Even though the combination was versatile—and quite amusing, the U.S. government decided the new substance wasn't a good substitute for synthetic rubber.

A few years later, the recipe for the stretch material fell into the hands of a businessperson, who saw the goo's potential—as a toy. The toymaker paid $147 for rights to the boric acid and silicone oil mixture. And in 1949 it was sold at toy stores for the first time. The material was packaged in a plastic egg and it took the U.S. by storm. Today, the acid and oil mixture comes in a multitude of colors and almost every child has played with it at some time.

The substance can be used for more than child's play. Its sticky consistency makes it good for cleaning computer keyboards and removing small specks of lint from fabrics.

People use it to make impressions of newspaper print or comics. Athletes strengthen their grips by grasping it over and over. Astronauts use it to anchor tools on spacecraft in zero gravity. All in all, a most *eggs-cellent* idea!

Research As a group, examine a sample of the colorful, sticky, stretch toy made of boric acid and silicone oil. Then brainstorm some practical—and impractical—uses for the substance.

Science online
For more information, visit
bookk.msscience.com/oops

Reviewing Main Ideas

Section 1 Matter

1. All matter is composed of tiny particles that are in constant motion.

2. In the solid state, the attractive force between particles holds them in place to vibrate.

3. Particles in the liquid state have defined volumes and are free to move about within the liquid.

Section 2 Changes of State

1. Thermal energy is the total energy of the particles in a sample of matter. Temperature is the average kinetic energy of the particles in a sample.

2. An object gains thermal energy when it changes from a solid to a liquid, or when it changes from a liquid to a gas.

3. An object loses thermal energy when it changes from a gas to a liquid, or when it changes from a liquid to a solid.

Section 3 Behavior of Fluids

1. Pressure is force divided by area.

2. Fluids exert a buoyant force in the upward direction on objects immersed in them.

3. An object will float in a fluid that is more dense than itself.

4. Pascal's principle states that pressure applied to a liquid is transmitted evenly throughout the liquid.

Visualizing Main Ideas

Copy and complete the following concept map on matter.

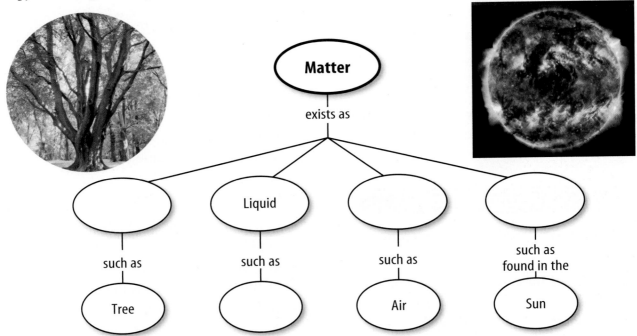

Matter

exists as

such as — Tree

Liquid — such as

such as — Air

such as found in the — Sun

Using Vocabulary

Archimedes' principle p. 59	melting p. 47
	Pascal's principle p. 60
buoyant force p. 58	pressure p. 54
condensation p. 51	solid p. 41
density p. 59	surface tension p. 43
freezing p. 49	temperature p. 46
gas p. 44	thermal energy p. 45
heat p. 46	vaporization p. 50
liquid p. 42	viscosity p. 43
matter p. 40	

Fill in the blanks with the correct vocabulary word.

1. A(n) _____ can change shape and volume.

2. A(n) _____ has a different shape but the same volume in any container.

3. _____ is thermal energy moving from one substance to another.

4. _____ is a measure of the average kinetic energy of the particles of a substance.

5. A substance changes from a gas to a liquid during the process of _____.

6. A liquid becomes a gas during _____.

7. _____ is mass divided by volume.

8. _____ is force divided by area.

9. _____ explains what happens when force is applied to a confined fluid.

Checking Concepts

Choose the word or phrase that best answers the question.

10. Which of these is a crystalline solid?
    - **A)** glass
    - **B)** sugar
    - **C)** rubber
    - **D)** plastic

11. Which description best describes a solid?
    - **A)** It has a definite shape and volume.
    - **B)** It has a definite shape but not a definite volume.
    - **C)** It adjusts to the shape of its container.
    - **D)** It can flow.

12. What property enables you to float a needle on water?
    - **A)** viscosity
    - **B)** temperature
    - **C)** surface tension
    - **D)** crystal structure

13. What happens to an object as its kinetic energy increases?
    - **A)** It holds more tightly to nearby objects.
    - **B)** Its mass increases.
    - **C)** Its particles move more slowly.
    - **D)** Its particles move faster.

14. During which process do particles of matter release energy?
    - **A)** melting
    - **B)** freezing
    - **C)** sublimation
    - **D)** boiling

15. How does water vapor in air form clouds?
    - **A)** melting
    - **B)** evaporation
    - **C)** condensation
    - **D)** sublimation

16. Which is a unit of pressure?
    - **A)** N
    - **B)** kg
    - **C)** g/cm^3
    - **D)** N/m^2

17. Which change results in an increase in gas pressure in a balloon?
    - **A)** decrease in temperature
    - **B)** decrease in volume
    - **C)** increase in volume
    - **D)** increase in altitude

18. In which case will an object float on a fluid?
    - **A)** Buoyant force is greater than weight.
    - **B)** Buoyant force is less than weight.
    - **C)** Buoyant force equals weight.
    - **D)** Buoyant force equals zero.

Science online bookk.msscience.com/vocabulary_puzzlemaker

Use the photo below to answer question 19.

19. In the photo above, the water in the small beaker was displaced when the golf ball was added to the large beaker. What principle does this show?
    A) Pascal's principle
    B) the principle of surface tension
    C) Archimedes' principle
    D) the principle of viscosity

20. Which is equal to the buoyant force on an object?
    A) volume of the object
    B) weight of the displaced fluid
    C) weight of object
    D) volume of fluid

Thinking Critically

21. **Explain** why steam causes more severe burns than boiling water.

22. **Explain** why a bathroom mirror becomes fogged while you take a shower.

23. **Form Operational Definitions** Write operational definitions that explain the properties of and differences among solids, liquids, and gases.

24. **Determine** A king's crown has a volume of 110 cm³ and a mass of 1,800 g. The density of gold is 19.3 g/cm³. Is the crown pure gold?

25. **Infer** Why do some balloons pop when they are left in sunlight for too long?

Performance Activities

26. **Storyboard** Create a visual-aid storyboard to show ice changing to steam. There should be a minimum of five frames.

Applying Math

Use the graph below to answer question 27.

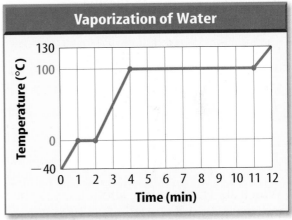

Vaporization of Water

27. **Explain** how this graph would change if a greater volume of water were heated. How would it stay the same?

Use the table below to answer question 28.

Water Pressure

Depth (m)	Pressure (atm)	Depth (m)	Pressure (atm)
0	1.0	100	11.0
25	3.5	125	13.5
50	6.0	150	16.0
75	8.5	175	18.5

28. **Make and Use Graphs** In July of 2001, Yasemin Dalkilic of Turkey dove to a depth of 105 m without any scuba equipment. Make a depth-pressure graph for the data above. Based on your graph, how does water pressure vary with depth? Note: The pressure at sea level, 101.3 kPa, is called one atmosphere (atm).

Part 1 Multiple Choice

Record your answers on the answer sheet provided by your teacher or on a sheet of paper.

1. In which state of matter do particles stay close together, yet are able to move past one another?
   A. solid
   B. gas
   C. liquid
   D. plasma

Use the illustration below to answer questions 2 and 3.

2. Which statement is true about the volume of the water displaced when the golf ball was dropped into the large beaker?
   A. It is equal to the volume of the golf ball.
   B. It is greater than the volume of the golf ball.
   C. It is less than the volume of the golf ball.
   D. It is twice the volume of a golf ball.

3. What do you know about the buoyant force on the golf ball?
   A. It is equal to the density of the water displaced.
   B. It is equal to the volume of the water displaced.
   C. It is less than the weight of the water displaced.
   D. It is equal to the weight of the water displaced.

4. What is the process called when a gas cools to form a liquid?
   A. condensation
   B. sublimation
   C. boiling
   D. freezing

5. Which of the following is an amorphous solid?
   A. diamond
   B. sugar
   C. glass
   D. sand

6. Which description best describes a liquid?
   A. It has a definite shape and volume.
   B. It has a definite volume but not a definite shape.
   C. It expands to fill the shape and volume of its container.
   D. It cannot flow.

7. During which processes do particles of matter absorb energy?
   A. freezing and boiling
   B. condensation and melting
   C. melting and vaporization
   D. sublimation and freezing

Use the illustration below to answer questions 8 and 9.

8. What happens as the piston moves down?
   A. The volume of the gas increases.
   B. The volume of the gas decreases.
   C. The gas particles collide less often.
   D. The pressure of the gas decreases.

9. What relationship between the volume and pressure of a gas does this illustrate?
   A. As volume decreases, pressure decreases.
   B. As volume decreases, pressure increases.
   C. As volume decreases, pressure remains the same.
   D. As the volume increases, pressure remains the same.

Part 2 | Short Response/Grid In

*Record your answers on the answer sheet
provided by your teacher or on a sheet of paper.*

10. A balloon filled with helium bursts in a
closed room. What space will the helium
occupy?

Use the illustration below to answer questions 11 and 12.

11. If the force exerted by the dancer is 510 N,
what is the pressure she exerts if the area is
335 cm^2 on the left and 37 cm^2 on the right?

12. Compare the pressure the dancer would
exert on the floor if she were wearing large
clown shoes to the photo on the left.

13. If a balloon is blown up and tied closed, air
is held inside it. What will happen to the
balloon if it is then pushed into hot water or
held over a heater? Why does this happen?

14. What is the relationship of heat and thermal energy?

15. Why are some insects able to move
around on the surface of a lake or pond?

16. How does the weight of a floating object
compare with the buoyant force acting on
the object?

17. What is the mass of an object that has a
density of 0.23 g/cm^3 and whose volume is
52 cm^3?

Part 3 | Open Ended

Record your answer on a sheet of paper.

18. Compare and contrast evaporation and
boiling.

Use the illustration below to answer questions 19 and 20.

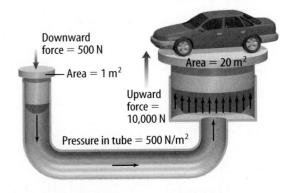

Downward
force = 500 N

Area = 1 m^2

Area = 20 m^2

Upward
force =
10,000 N

Pressure in tube = 500 N/m^2

19. Name and explain the principle that is
used in lifting the car.

20. Explain what would happen if you dou-
bled the area of the piston on the right
side of the hydraulic system.

21. Explain why a woman might put dents in
a wood floor when walking across it in
high-heeled shoes, but not when wearing
flat sandals.

22. Explain why the tires on a car might
become flattened on the bottom after sit-
ting outside in very cold weather.

23. Compare the arrangement and movement
of the particles in a solid, a liquid, and a gas.

24. Explain why the water in a lake is much
cooler than the sand on the beach around
it on a sunny summer day.

Test-Taking Tip

Show Your Work For open-ended questions, show all of your
work and any calculations on your answer sheet.

Hint: In question 20, the pressure in the tube does not change.

Properties and Changes of Matter

The BIG Idea

Matter is classified by physical and chemical properties and changes.

SECTION 1
Physical and Chemical Properties
Main Idea All matter has physical and chemical properties.

SECTION 2
Physical and Chemical Changes
Main Idea Matter undergoes physical and chemical changes.

Volcanic Eruptions

At very high temperatures deep within Earth, solid rock melts. One of the properties of rock is its state—solid, liquid, or gas. As lava changes from a liquid to a solid, what happens to its properties? In this chapter, you will learn about physical and chemical properties and changes of matter.

Science Journal Think about what happens when you crack a glow stick. What types of changes are you observing?

Start-Up Activities

The Changing Face of a Volcano

When a volcano erupts, it spews lava and gases. Lava is hot, melted rock from deep within the Earth. After it reaches the Earth's surface, the lava cools and hardens into solid rock. The minerals and gases within the lava, as well as the rate at which it cools, determine the characteristics of the resulting rocks. In this lab, you will compare two types of volcanic rock.

1. Obtain similar-sized samples of the rocks obsidian (ub SIH dee un) and pumice (PUH mus) from your teacher.

2. Compare the colors of the two rocks.

3. Decide which sample is heavier.

4. Look at the surfaces of the two rocks. How are the surfaces different?

5. Place each rock in water and observe.

6. **Think Critically** What characteristics are different about these rocks? In your Science Journal, make a table that compares your observations.

Science nline | Preview this chapter's content and activities at bookk.msscience.com

Changes of Matter Make the following Foldable to help you organize your thoughts about properties and changes.

STEP 1 Fold a sheet of paper in half lengthwise. Make the back edge about 1.25 cm longer than the front edge.

STEP 2 Fold in half, then fold in half again to make three folds.

STEP 3 Unfold and cut only the top layer along the three folds to make four tabs.

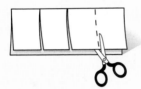

STEP 4 Label the tabs as shown.

Find Main Ideas As you read the chapter, write information about matter's physical and chemical properties and changes.

Get Ready to Read

Identify the Main Idea

① Learn It! Main ideas are the most important ideas in a paragraph, section, or chapter. Supporting details are facts or examples that explain the main idea. Understanding the main idea allows you to grasp the whole picture.

② Picture It! Read the following paragraph. Draw a graphic organizer like the one below to show the main idea and supporting details.

> Some physical properties describe the appearance of matter. You can detect many of these properties with your senses. For example, you can see the color and shape of the ride at the fair. You can also touch it to feel its texture. You can smell the odor or taste the flavor of some matter.
>
> —*from page 73*

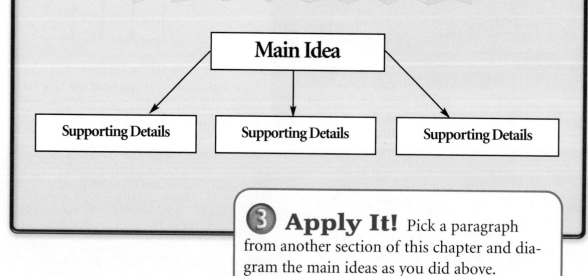

③ Apply It! Pick a paragraph from another section of this chapter and diagram the main ideas as you did above.

Target Your Reading

Reading Tip

To help understand a difficult paragraph, identify the main idea and then identify the supporting details.

Use this to focus on the main ideas as you read the chapter.

① **Before you read** the chapter, respond to the statements below on your worksheet or on a numbered sheet of paper.
- Write an **A** if you **agree** with the statement.
- Write a **D** if you **disagree** with the statement.

② **After you read** the chapter, look back to this page to see if you've changed your mind about any of the statements.
- If any of your answers changed, explain why.
- Change any false statements into true statements.
- Use your revised statements as a study guide.

Science Online
Print out a worksheet of this page at
bookk.msscience.com

Before You Read A or D		Statement	After You Read A or D
	1	A physical property can be observed with your senses.	
	2	State, density, melting point, and boiling point are chemical properties.	
	3	The ability to attract some metal objects is a physical property of lodestone.	
	4	Mass, weight, and volume are size-dependent properties.	
	5	A change of state is a chemical change.	
	6	A color change is a sign of a chemical change.	
	7	The release of energy is a sign of a physical change.	
	8	If the chemical composition of a substance changes, it has undergone a chemical change.	
	9	When wood burns, it undergoes a chemical change.	

Physical and Chemical Properties

as you read

What You'll Learn
- **Identify** physical and chemical properties of matter.
- **Classify** objects based on physical properties.

Why It's Important
Understanding the different properties of matter will help you to better describe the world around you.

🔍 **Review Vocabulary**
matter: anything that has mass and takes up space

New Vocabulary
- physical property
- chemical property

Physical Properties

It's a busy day at the state fair as you and your classmates navigate your way through the crowd. While you follow your teacher, you can't help but notice the many sights and sounds that surround you. Eventually, you fall behind the group as you spot the most amazing ride you have ever seen. You inspect it from one end to the other. How will you describe it to the group when you catch up to them? What features will you use in your description?

Perhaps you will mention that the ride is large, blue, and made of wood. These features are all physical properties, or characteristics, of the ride. A **physical property** is a characteristic that you can observe without changing or trying to change the composition of the substance. How something looks, smells, sounds, or tastes are all examples of physical properties. In **Figure 1** you can describe and differentiate all types of matter by observing their properties.

✔ **Reading Check** *What is a physical property of matter?*

Figure 1 All matter can be described by physical properties that can be observed using the five senses.
Identify *the types of matter you think you could see, hear, taste, touch, and smell at the fair.*

Using Your Senses Some physical properties describe the appearance of matter. You can detect many of these properties with your senses. For example, you can see the color and shape of the ride at the fair. You can also touch it to feel its texture. You can smell the odor or taste the flavor of some matter. (You should never taste anything in the laboratory.) Consider the physical properties of the items in **Figure 2.**

State To describe a sample of matter, you need to identify its state. Is the ride a solid, a liquid, or a gas? This property, known as the state of matter, is another physical property that you can observe. The ride, your chair, a book, and a pen are examples of matter in the solid state. Milk, gasoline, and vegetable oil are examples of matter in the liquid state. The helium in a balloon, air in a tire, and neon in a sign are examples of matter in the gas state. You can see examples of solids, liquids, and gases in **Figure 3.**

Perhaps you are most familiar with the three states of water. You can drink or swim in liquid water. You use the solid state of water, which is ice, when you put ice cubes in a drink or skate on a frozen lake. Although you can't see it, water in the gas state is all around you in the air.

Figure 2 Some matter has a characteristic color, such as this sulfur pile. You can use a characteristic smell or taste to identify these fruits. Even if you didn't see it, you could probably identify this sponge by feeling its texture.

Figure 3 The state of a sample of matter is an important physical property.

This rock formation is in the solid state.

The oil flowing out of a bottle is in the liquid state.

This colorful sign uses the element neon, which is generally found in the gaseous state.

Mini LAB

Measuring Properties

Procedure 🥽 🧤 🧹

1. Measure the mass of a **10-mL graduated cylinder.**
2. Fill the graduated cylinder with **water** to the 10-mL mark and remeasure the mass of the graduated cylinder with the water.
3. Determine the mass of the water by subtracting the mass of the graduated cylinder from the mass of the graduated cylinder and water.
4. Determine the density of water by dividing the mass of the water by the volume of the water.

Analysis

1. Why did you need to measure the mass of the empty graduated cylinder?
2. How would your calculated density be affected if you added more than 10 mL of water?

Figure 4 A spring scale is used to measure an object's weight.

Size-Dependent Properties Some physical properties depend on the size of the object. Suppose you need to move a box. The size of the box would be important in deciding if you need to use your backpack or a truck. You begin by measuring the width, height, and depth of the box. If you multiply them together, you calculate the box's volume. The volume of an object is the amount of space it occupies.

Another physical property that depends on size is mass. Recall that the mass of an object is a measurement of how much matter it contains. A bowling ball has more mass than a basketball. Weight is a measurement of force. Weight depends on the mass of the object and on gravity. If you were to travel to other planets, your weight would change but your size and mass would not. Weight is measured using a spring scale like the one in **Figure 4.**

Size-Independent Properties Another physical property, density, does not depend on the size of an object. Density measures the amount of mass in a given volume. To calculate the density of an object, divide its mass by its volume. The density of water is the same in a glass as it is in a tub. The density of an object will change, however, if the mass changes and the volume remains the same. Another property, solubility, also does not depend on size. Solubility is the number of grams of one substance that will dissolve in 100 g of another substance at a given temperature. The amount of drink mix that can be dissolved in 100 g of water is the same in a pitcher as it is when it is poured into a glass. Size-dependent and independent properties are shown in **Table 1.**

Melting and Boiling Point Melting and boiling point also do not depend upon an object's size. The temperature at which a solid changes into a liquid is called its melting point. The temperature at which a liquid changes into a gas is called its boiling point. The melting and boiling points of several substances, along with some of their other physical properties, are shown in **Table 2.**

Table 1 Properties of Matter		
Physical Properties		
Dependent on sample size	mass, weight, volume	
Independent of sample size	density, melting/boiling point, solubility, ability to attract a magnet, state of matter, color	

Table 2 Physical Properties of Several Substances

Substance	State	Density (g/cm³)	Melting point (°C)	Boiling point (°C)	Solubility in cold water (g/100 mL)
Ammonia	gas	0.7710	-78	-33	89.9
Bromine	liquid	3.12	-7	59	4.17
Calcium carbonate	solid	2.71	898	1,339	0.0014
Iodine	solid	4.93	113.5	184	0.029
Potassium hydroxide	solid	2.044	360	1,322	107
Sodium chloride	solid	2.17	801	1,413	35.7
Water	liquid	1	0	100	—

Magnetic Properties Some matter can be described by the specific way in which it behaves. For example, some materials pull iron toward them. These materials are said to be magnetic. The lodestone in **Figure 5** is a rock that is naturally magnetic.

Other materials can be made into magnets. You might have magnets on your refrigerator or locker at school. The door of your refrigerator also has a magnet within it that holds the door shut tightly.

Reading Check *What are some examples of physical properties of matter?*

Figure 5 This lodestone attracts certain metals to it. Lodestone is a natural magnet.

Mini LAB

Identifying an Unknown Substance

Procedure
1. Obtain data from your teacher (mass, volume, solubility, melting or boiling point) for an unknown substance(s).
2. Calculate density and solubility in units of g/100 mL for your unknown substance(s).
3. Using Table 2 and the information you have, identify your unknown substance(s).

Analysis
1. Describe the procedure used to determine the density of your unknown substance(s).
2. Identify three characteristics of your substance(s).
3. Explain how the solubility of your substance would be affected if the water was hot.

Figure 6 Notice the difference between the new matches and the matches that have been burned. The ability to burn is a chemical property of matter.

Topic: Measuring Matter
Visit bookk.msscience.com for Web links to information about methods of measuring matter.

Activity Find an object around the house. Use two methods of measuring matter to describe it.

Chemical Properties

Some properties of matter cannot be identified just by looking at a sample. For example, nothing happens if you look at the matches in the first picture. But if someone strikes the matches on a hard, rough surface they will burn, as shown in the second picture. The ability to burn is a chemical property. A **chemical property** is a characteristic that cannot be observed without altering the substance. As you can see in the last picture, the matches are permanently changed after they are burned. Therefore this property can be observed only by changing the composition of the match. Another way to define a chemical property, then, is the ability of a substance to undergo a change that alters its identity. You will learn more about changes in matter in the following section.

section 1 review

Summary

Physical Properties
- Matter exists in solid, liquid, and gaseous states.
- Volume, mass, and weight are size-dependent properties.
- Properties such as density, solubility, boiling and melting points, and ability to attract a magnet are size-independent.
- Density relates the mass of an object to its volume.

Chemical Properties
- Chemical properties have characteristics that cannot be observed without altering the identity of the substance.

Self Check

1. **Infer** How are your senses important for identifying physical properties of matter?
2. **Describe** the physical properties of a baseball.
3. **Think Critically** Explain why solubility is a size-independent physical property.
4. **Compare and Contrast** How do chemical and physical properties differ?

Applying Math

5. **Solve One-Step Equations** The volume of a bucket is 5 L and you are using a cup with a volume of 50 mL. How many cupfuls will you need to fill the bucket? Hint: 1 L = 1,000 mL

Finding the Difference

⊙ *Real-World Question*

You can identify an unknown object by comparing its physical and chemical properties to the properties of identified objects.

Goals

- ■ **Identify** the physical properties of objects.
- ■ **Compare and contrast** the properties.
- ■ **Categorize** the objects based on their properties.

Materials

meterstick	rock
spring scale	plant or flower
block of wood	soil
metal bar or metal ruler	sand
plastic bin	apple (or other fruit)
drinking glass	vegetable
water	slice of bread
rubber ball	dry cereal
paper	egg
carpet	feather
magnet	

Safety Precautions

⊙ *Procedure*

1. List at least six properties that you will observe, measure, or calculate for each object. Describe how to determine each property.

2. In your Science Journal, create a data table with a column for each property and rows for the objects.

3. Complete your table by determining the properties for each object.

⊙ *Conclude and Apply*

1. **Describe** Which properties were you able to observe easily? Which required making measurements? Which required calculations?

2. **Compare and contrast** the objects based on the information in your table.

3. **Draw Conclusions** Choose a set of categories and group your objects into those categories. Some examples of categories are large/medium/small, heavy/moderate/light, bright/moderate/dull, solid/liquid/gas, etc. Were the categories you chose useful for grouping your objects? Why or why not?

𝒞ommunicating
Your Data

Compare your results with those of other students in your class. **Discuss** the properties of objects that different groups included on their tables. Make a large table including all of the objects that students in the class studied.

Physical and Chemical Changes

What You'll Learn

■ **Compare** several physical and chemical changes.
■ **Identify** examples of physical and chemical changes.

Why It's Important

From modeling clay to watching the leaves turn colors, physical and chemical changes are all around us.

🔍 **Review Vocabulary**

solubility: the amount of a substance that will dissolve in a given amount of another substance

New Vocabulary

● physical change
● vaporization
● condensation
● sublimation
● deposition
● chemical change
● law of conservation of mass

Physical Changes

What happens when the artist turns the lump of clay shown in **Figure 7** into bowls and other shapes? The composition of the clay does not change. Its appearance, however, changes dramatically. The change from a lump of clay to different shapes is a physical change. A **physical change** is one in which the form or appearance of matter changes, but not its composition. The lake in **Figure 7** also experiences a physical change. Although the water changes state due to a change in temperature, it is still made of the elements hydrogen and oxygen.

Changing Shape Have you ever crumpled a sheet of paper into a ball? If so, you caused physical change. Whether it exists as one flat sheet or a crumpled ball, the matter is still paper. Similarly, if you cut fruit into pieces to make a fruit salad, you do not change the composition of the fruit. You change only its form. Generally, whenever you cut, tear, grind, or bend matter, you are causing a physical change.

Figure 7 Although each sample looks quite different after it experiences a change, the composition of the matter remains the same. These changes are examples of physical changes.

Dissolving What type of change occurs when you add sugar to iced tea, as shown in **Figure 8?** Although the sugar seems to disappear, it does not. Instead, the sugar dissolves. When this happens, the particles of sugar spread out in the liquid. The composition of the sugar stays the same, which is why the iced tea tastes sweet. Only the form of the sugar has changed.

Figure 8 Physical changes are occurring constantly. The sugar blending into the iced tea is an example of a physical change. **Define** *What is a physical change?*

Changing State Another common physical change occurs when matter changes from one state to another. When an ice cube melts, for example, it becomes liquid water. The solid ice and the liquid water have the same composition. The only difference is the form.

Matter can change from any state to another. Freezing is the opposite of melting. During freezing, a liquid changes into a solid. A liquid also can change into a gas. This process is known as **vaporization.** During the reverse process, called **condensation,** a gas changes into a liquid. **Figure 9** summarizes these changes.

In some cases, matter changes between the solid and gas states without ever becoming a liquid. The process in which a solid changes directly into a gas is called **sublimation.** The opposite process, in which a gas changes into a solid, is called **deposition.**

Figure 9 Look at the photographs below to identify the different physical changes that bromine undergoes as it changes from one state to another.

Solid state

Gas state

Liquid state

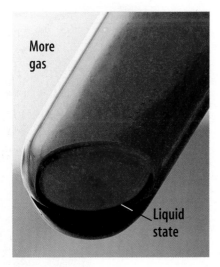

More gas

Liquid state

Chemical Changes

It's the Fourth of July in New York City. Brilliant fireworks are exploding in the night sky. When you look at fireworks, such as these in **Figure 10,** you see dazzling sparkles of red and white trickle down in all directions. The explosion of fireworks is an example of a chemical change. During a **chemical change,** substances are changed into different substances. In other words, the composition of the substance changes.

You are familiar with another chemical change if you have ever left your bicycle out in the rain. After awhile, a small chip in the paint leads to an area of a reddish, powdery substance. This substance is rust. When iron in steel is exposed to oxygen and water in air, iron and oxygen atoms combine to form the principle component in rust. In a similar way, silver coins tarnish when exposed to air. These chemical changes are shown in **Figure 11.**

Figure 10 These brilliant fireworks result from chemical changes.
Define *What is a chemical change?*

 **Reading Check** *How is a chemical change different from a physical change?*

Figure 11 Each of these examples shows the results of a chemical change. In each case, the substances that are present after the change are different from those that were present before the change.

Figure 12 In the fall, the chlorophyll in this tree's leaves undergoes a chemical change into colorless chemicals. This allows the red pigment to be seen.

Signs of Chemical Changes

Physical changes are relatively easy to identify. If only the form of a substance changes, you have observed a physical change. How can you tell whether a change is a chemical change? If you think you are unfamiliar with chemical changes, think again.

INTEGRATE Life Science You have witnessed a spectacular chemical change if you have seen the leaves on a tree change from green to bright yellow, red, or orange. But, it is not a change from a green pigment to a red pigment, as you might think. Pigments are chemicals that give leaves their color. In **Figure 12,** the green pigment that you see during the summer is chlorophyll (KLOHR uh fihl). In autumn, however, changes in temperature and rainfall amounts cause trees to stop producing chlorophyll. The chlorophyll already in the leaves undergoes a chemical change into colorless chemicals. Where do the bright fall colors come from? The pigments that produce fall colors have been present in the leaves all along. However, in the summer, chlorophyll is present in large enough amounts to mask these pigments. In the fall, when chlorophyll production stops, the bright pigments become visible.

Color Perhaps you have found that a half-eaten apple turns brown. The reason is that a chemical change occurs when the apple is exposed to air. Maybe you have toasted a marshmallow or a slice of bread and watched them turn black. In each case, the color of the food changes as it is cooked because a chemical change occurs.

Science Online

Topic: Recognizing Chemical Changes
Visit bookk.msscience.com for Web links to information about how chemical equations can be used to model chemical changes.

Activity Describe the chemical reactions that are involved in making and baking a yeast bread.

Mini LAB

Comparing Changes

Procedure

1. Separate a piece of **fine steel wool** into two halves.
2. Dip one half in **tap water.**
3. Place each piece of steel wool on a separate **paper plate** and let them sit overnight.

Analysis

1. Did you observe any changes in the steel wool? If so, describe them.
2. If you observed changes, were they physical or chemical? How do you know?

Try at Home

Figure 13 Cake batter undergoes a chemical change as it absorbs energy during cooking.

Energy Another sign of a chemical change is the release or gain of energy by an object. Many substances must absorb energy in order to undergo a chemical change. For example, energy is absorbed during the chemical changes involved in cooking. When you bake a cake or make pancakes, energy is absorbed by the batter as it changes from a runny mix into what you see in **Figure 13.**

Another chemical change in which a substance absorbs energy occurs during the production of cement. This process begins with the heating of limestone. Ordinarily, limestone will remain unchanged for centuries. But when it absorbs energy during heating, it undergoes a chemical change in which it turns into lime and carbon dioxide.

Energy also can be released during a chemical change. The fireworks you read about earlier released energy in the form of light that you can see. As shown in **Figure 14,** a chemical change within a firefly releases energy in the form of light. Fuel burned in the camping stove releases energy you see as light and feel as heat. You also can see that energy is released when sodium and chlorine are combined and ignited in the last picture. During this chemical change, the original substances change into sodium chloride, which is ordinary table salt.

Figure 14 Energy is released when a firefly glows, when fuel is burned in a camping stove, and when sodium and chlorine undergo a chemical change to form table salt.

Odor It takes only one experience with a rotten egg to learn that they smell much different than fresh eggs. When eggs and other foods spoil, they undergo chemical change. The change in odor is a clue to the chemical change. This clue can save lives. When you smell an odd odor in foods, such as chicken, pork, or mayonnaise, you know that the food has undergone a chemical change. You can use this clue to avoid eating spoiled food and protect yourself from becoming ill.

Gases or Solids Look at the antacid tablet in **Figure 15.** You can produce similar bubbles if you pour vinegar on baking soda. The formation of a gas is a clue to a chemical change. What other products undergo chemical changes and produce bubbles?

Figure 15 also shows another clue to a chemical change—the formation of a solid. A solid that separates out of a solution during a chemical change is called a precipitate. The precipitate in the photograph forms when a solution containing sodium iodide is mixed with a solution containing lead nitrate.

INTEGRATE Astronomy

Meteoroid A meteoroid is a chunk of metal or stone in space. Every day, meteoroids enter Earth's atmosphere. When this happens, the meteoroid burns as a result of friction with gases in the atmosphere. It is then referred to as a meteor, or shooting star. The burning produces streaks of light. The burning is an example of a chemical change. In your Science Journal, infer why most meteoroids never reach Earth's surface.

Figure 15 The bubbles of gas formed when this antacid tablet is dropped into water indicate a chemical change. The solid forming from two liquids is another sign that a chemical change has taken place.

Figure 16 As wood burns, it turns into a pile of ashes and gases that rise into the air.
Determine *Can you turn ashes back into wood?*

Not Easily Reversed How do physical and chemical changes differ from one another? Think about ice for a moment. After solid ice melts into liquid water, it can refreeze into solid ice if the temperature drops enough. Freezing and melting are physical changes. The substances produced during a chemical change cannot be changed back into the original substances by physical means. For example, the wood in **Figure 16** changes into ashes and gases that are released into the air. After wood is burned, it cannot be restored to its original form as a log.

Think about a few of the chemical changes you just read about to see if this holds true. An antacid tablet cannot be restored to its original form after being dropped in water. Rotten eggs cannot be made fresh again, and pancakes cannot be turned back into batter. The substances that existed before the chemical change no longer exist.

> ✓ **Reading Check** *What signs indicate a chemical change?*

Applying Math Solve for an Unknown

CONVERTING TEMPERATURES Fahrenheit is a non-SI temperature scale. Because it is used so often, it is useful to be able to convert from Fahrenheit to Celsius. The equation that relates Celsius degrees to Fahrenheit degrees is: (°C × 1.8) + 32 = °F. What is 15°F on the Celsius scale?

Solution

1 *This is what you know:*
- temperature = 15°F
- (°C × 1.8) + 32 = °F

2 *This is what you need to find out:*
- temperature in degrees Celsius

3 *This is the procedure you need to use:*
- (°C × 1.8) + 32 = °F
- °C = (°F − 32)/1.8
- °C = (15 − 32)/1.8 = −9.4°C

4 *Check your answer:*
Substitute the Celsius temperature into the original equation. Did you calculate the Fahrenheit temperature that was given?

Practice Problems

1. Water is being heated on the stove at 156°F. What is this temperature on the Celsius scale?

2. The boiling point of ethylene glycol is 199°C. What is the temperature on the Fahrenheit scale?

 For more practice, visit bookk.msscience.com/math_practice

Chemical Versus Physical Change

Now you have learned about many different physical and chemical changes. You have read about several characteristics that you can use to distinguish between physical and chemical changes. The most important point for you to remember is that in a physical change, the composition of a substance does not change and in a chemical change, the composition of a substance does change. When a substance undergoes a physical change, only its form changes. In a chemical change, both form and composition change.

When the wood and copper in **Figure 17** undergo physical changes, the original wood and copper still remain after the change. When a substance undergoes a chemical change, however, the original substance is no longer present after the change. Instead, different substances are produced during the chemical change. When the wood and copper in **Figure 17** undergo chemical changes, wood and copper have changed into new substances with new physical and chemical properties.

Physical and chemical changes are used to recycle or reuse certain materials. **Figure 18** discusses the importance of some of these changes in recycling.

Figure 17 When a substance undergoes a physical change, its composition stays the same. When a substance undergoes a chemical change, it is changed into different substances.

Chemical change

Chemical change

Physical change

Physical change

Figure 18

Recycling is a way to separate wastes into their component parts and then reuse those components in new products. In order to be recycled, wastes need to be physically—and sometimes chemically—changed. The average junked automobile contains about 62 percent iron and steel, 28 percent other materials such as aluminum, copper, and lead, and 10 percent rubber, plastics, and various materials.

▼ After being crushed and flattened, car bodies are chopped into small pieces. Metals are separated from other materials using physical processes. Some metals are separated using powerful magnets. Others are separated by hand.

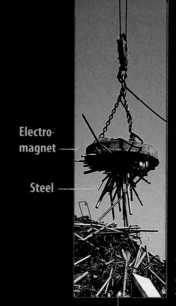

Electromagnet —

Steel —

◀ Rubber tires can be shredded and added to asphalt pavement and playground surfaces. New recycling processes make it possible to supercool tires to a temperature at which the rubber is shattered like glass. A magnet can then draw out steel from the tires and other parts of the car.

◀ Glass can be pulverized and used in asphalt pavement, new glass, and even artwork. This sculpture, named *Groundswell,* was created by artist Maya Lin using windshield glass.

▲ Some plastics can be melted and formed into new products. Others are ground up or shredded and used as fillers or insulating materials.

Conservation of Mass

During a chemical change, the form or the composition of the matter changes. The particles within the matter rearrange to form new substances, but they are not destroyed and new particles are not created. The number and type of particles remains the same. As a result, the total mass of the matter is the same before and after a physical or chemical change. This is known as the **law of conservation of mass.**

This law can sometimes be difficult to believe, especially when the materials remaining after a chemical change might look quite different from those before it. In many chemical changes in which mass seems to be gained or lost, the difference is often due to a gas being given off or taken in. When the candle burns in **Figure 19,** gases in the air combine with the candle wax. New gases are formed that go into the air. The mass of the wax, which is burned and the gases that combine with the wax equal the mass of the gases produced by burning.

The scientist who first performed the careful experiments necessary to prove that mass is conserved was Antoine Lavoisier (AN twan • luh VWAH see ay) in the eighteenth century. It was Lavoisier who recognized that the mass of gases that are given off or taken from the air during chemical changes account for any differences in mass.

Figure 19 The candle lost mass when it was burned. The mass lost by the candle combined with the gases in the air to form new substances. As a result, mass was not created or destroyed.

section 2 review

Summary

Physical Changes
- The form of matter, its shape or state, is altered during a physical change.
- The composition of matter remains the same.

Chemical Changes
- Both form and composition of matter are altered during a chemical change.
- Some signs of a chemical change are altered color, energy, odor, and formation of a gas or solid.
- Chemical changes are not easily reversed.

Conservation of Mass
- The total mass of the matter is the same before and after a physical or chemical change.

Self Check

1. **List** five physical changes that you can observe in your home.
2. **Determine** what kind of change occurs on the surface of bread when it is toasted.
3. **Infer** How is mass conserved during a chemical change?
4. **Think Critically** A log is reduced to a small pile of ash when it burns. Explain the difference in mass between the log and the ash.

Applying Math

5. **Solve One-Step Equations** Magnesium and oxygen undergo a chemical change to form magnesium oxide. How many grams of magnesium oxide will be produced when 0.486 g of oxygen completely react with 0.738 g of magnesium?

BATTLE OF THE TOOTHPASTES

Goals

■ **Observe** how tooth-paste helps prevent tooth decay.

■ **Design** an experiment to test the effectiveness of various types and brands of toothpaste.

Possible Materials

3 or 4 different brands and types of toothpaste
drinking glasses or bowls
hard-boiled eggs
concentrated lemon juice
apple juice
water
artist's paint brush

Safety Precautions

◉ Real-World Question

Your teeth are made of a compound called hydrox-yapatite (hi DRAHK see A puh tite). The sodium fluoride in toothpaste undergoes a chemical reaction with hydroxyapatite to form a new compound on the surface of your teeth. This compound resists food acids that cause tooth decay, another chemical change. In this lab, you will design an experiment to test the effectiveness of different toothpaste brands. The compound found in your teeth is similar to the mineral compound found in eggshells. Treating hard-boiled eggs with toothpaste is similar to brushing your teeth with toothpaste. Soaking the eggs in food acids such as vinegar for several days will produce similar conditions as eating foods, which contain acids that will produce a chemical change in your teeth, for several months.

◉ Form a Hypothesis

Form a hypothesis about the effectiveness of different brands of toothpaste.

◉ Test Your Hypothesis

Make a Plan

1. **Describe** how you will use the materials to test the toothpaste.

2. **List** the steps you will follow to test your hypothesis.

3. **Decide** on the length of time that you will conduct your experiment.

4. **Identify** the control and variables you will use in your experiment.

5. **Create** a data table in your Science Journal to record your observations, measurements, and results.

6. **Describe** how you will measure the amount of protection each toothpaste brand provides.

Follow Your Plan

1. Make sure your teacher approves your plan before you start.

2. **Conduct** your experiment as planned. Be sure to follow all proper safety precautions.

3. **Record** your observations in your data table.

▶ *Analyze Your Data*

1. **Compare** the untreated eggshells with the shells you treated with toothpaste.

2. **Compare** the condition of the eggshells you treated with different brands of toothpaste.

3. **Compare** the condition of the eggshells soaked in lemon juice and in apple juice.

4. **Identify** unintended variables you discovered in your experiment that might have influenced the results.

▶ *Conclude and Apply*

1. **Identify** Did the results support your hypothesis? Describe the strengths and weaknesses of your hypothesis.

2. **Explain** why the eggshells treated with toothpaste were better-protected than the untreated eggshells.

3. **Identify** which brands of toothpaste, if any, best protected the eggshells from decay.

4. **Evaluate** the scientific explanation for why adding fluoride to toothpaste and drinking water prevents tooth decay.

5. **Predict** what would happen to your protected eggs if you left them in the food acids for several weeks.

6. **Infer** why it is a good idea to brush with fluoride toothpaste.

*C*ommunicating
Your Data

Compare your results with the results of your classmates. **Create** a poster advertising the benefits of fluoride toothpaste.

SCIENCE Stats

Strange Changes

Did you know...

... A hair colorist is also a chemist!
Colorists use hydrogen peroxide and ammonia to swell and open the cuticle-like shafts on your hair. Once these are open, the chemicals in hair dye can get into your natural pigment molecules and chemically change your hair color. The first safe commercial hair color was created in 1909 in France.

... Americans consume about 175 million kg of sauerkraut each year. During the production of sauerkraut, bacteria produce lactic acid. The acid chemically breaks down the material in the cabbage, making it translucent and tangy.

Applying Math There are 275 million people in the United States. Calculate the average amount of sauerkraut consumed by each person in the United States in one year.

... More than 450,000 metric tons of plastic packaging are recycled each year in the U.S. Discarded plastics undergo physical changes including melting and shredding. They are then converted into flakes or pellets, which are used to make new products. Recycled plastic is used to make clothes, furniture, carpets, and even lumber.

Projected Recycling Rates by Material, 2000		
Material	**1995 Recycling**	**Proj. Recycling**
Paper/Paperboard	40.0%	43 to 46%
Glass	24.5%	27 to 36%
Ferrous metal	36.5%	42 to 55%
Aluminum	34.6%	46 to 48%
Plastics	5.3%	7 to 10%
Yard waste	30.3%	40 to 50%
Total Materials	27.0%	30 to 35%

Find Out About It

Every time you cook, you make physical and chemical changes to food. Visit bookk.msscience.com/science_stats or to your local or school library to find out what chemical or physical changes take place when cooking ingredients are heated or cooled.

90 ◆ K CHAPTER 3 Properties and Changes of Matter

Reviewing Main Ideas

Section 1 Physical and Chemical Properties

1. Matter can be described by its characteristics, or properties, and can exist in different states—solid, liquid, or gas.

2. A physical property is a characteristic that can be observed without altering the composition of the sample.

3. Physical properties include color, shape, smell, taste, and texture, as well as measurable quantities such as mass, volume, density, melting point, and boiling point.

4. A chemical property is a characteristic that cannot be observed without changing what the sample is made of.

Section 2 Physical and Chemical Changes

1. During a physical change, the composition of matter stays the same but the appearance changes in some way.

2. Physical changes occur when matter changes from one state to another.

3. A chemical change occurs when the composition of matter changes.

4. Signs of chemical change include changes in energy, color, odor, or the production of gases or solids.

5. According to the law of conservation of mass, mass cannot be created or destroyed.

Visualizing Main Ideas

Copy and complete the following concept map on matter.

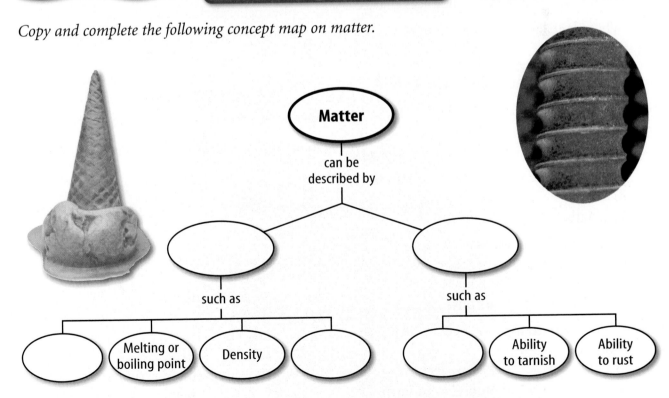

chapter 3 Review

Using Vocabulary

chemical change p. 80
chemical property p. 76
condensation p. 79
deposition p. 79
law of conservation
   of mass p. 87

physical change p. 78
physical property p. 72
sublimation p. 79
vaporization p. 79

Use what you know about the vocabulary words to answer the following questions. Use complete sentences.

1. Why is color a physical property?

2. What is a physical property that does not change with the amount of matter?

3. What happens during a physical change?

4. What type of change is a change of state?

5. What happens during a chemical change?

6. What are three clues that a chemical change has occurred?

7. What is an example of a chemical change?

8. What is the law of conservation of mass?

Checking Concepts

Choose the word or phrase that best answers the question.

9. What changes when the mass of an object increases while volume stays the same?
   A) color
   B) length
   C) density
   D) height

10. What word best describes the type of materials that attract iron?
   A) magnetic
   B) chemical
   C) mass
   D) physical

11. Which is an example of a chemical property?
   A) color
   B) mass
   C) density
   D) ability to burn

12. Which is an example of a physical change?
   A) metal rusting
   B) silver tarnishing
   C) water boiling
   D) paper burning

13. What characteristic best describes what happens during a physical change?
   A) composition changes
   B) composition stays the same
   C) form stays the same
   D) mass is lost

14. Which is an example of a chemical change?
   A) water freezes
   B) wood is carved
   C) bread is baked
   D) wire is bent

15. Which is NOT a clue that could indicate a chemical change?
   A) change in color
   B) change in shape
   C) change in energy
   D) change in odor

16. What property stays the same during physical and chemical changes?
   A) density
   B) shape
   C) mass
   D) arrangement of particles

Use the illustration below to answer question 17.

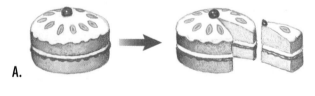
A.

B.

17. Which is an example of a physical change and which is a chemical change?

Science Online bookk.msscience.com/vocabulary_puzzlemaker

Thinking Critically

18. Draw Conclusions When asked to give the physical properties of a painting, your friend says the painting is beautiful. Why isn't this description a true scientific property?

19. Draw Conclusions You are told that a sample of matter gives off energy as it changes. Can you conclude which type of change occurred? Why or why not?

20. Describe what happens to mass during chemical and physical changes. Explain.

21. Classify Decide whether the following properties are physical or chemical.
- **a.** Sugar can change into alcohol.
- **b.** Iron can rust.
- **c.** Alcohol can vaporize.
- **d.** Paper can burn.
- **e.** Sugar can dissolve.

Use the table below to answer question 22 and 23.

Physical Properties

Substance	Melting Point (°C)	Density (g/cm³)
Benzoic acid	122.1	1.075
Sucrose	185.0	1.581
Methane	−182.0	0.466
Urea	135.0	1.323

22. Determine A scientist has a sample of a substance with a mass of 1.4 g and a volume of 3.0 mL. According to the table above, which substance might it be?

23. Conclude Using the table above, which substance would take the longest time to melt? Explain your reasoning.

24. Determine A jeweler bends gold into a beautiful ring. What type of change is this? Explain.

25. Compare and Contrast Relate such human characteristics as hair and eye color and height and weight to physical properties of matter. Relate human behavior to chemical properties. Think about how you observe these properties.

Performance Activities

26. Write a Story Write a story describing an event that you have experienced. Then go back through the story and circle any physical or chemical properties you mentioned. Underline any physical or chemical changes you included.

Applying Math

27. Brick Volume What is the volume of a brick that is 20 cm long, 10 cm wide, and 3 cm high?

28. Density of an Object What is the density of an object with a mass of 50 g and a volume of 5 cm³?

Use the table below to answer question 29.

Mineral Samples

Sample	Mass	Volume
A	96.5 g	5 cm³
B	38.6 g	4 cm³

29. Density of Gold The density of gold is 19.3 g/cm³. Which sample is the gold?

30. Ammonia Solubility 89.9 g of ammonia will dissolve in 100 mL of cold water. How much ammonia is needed to dissolve in 1.5 L of water?

Part 1 | Multiple Choice

Record your answers on the answer sheet provided by your teacher or on a sheet of paper.

Use the photograph below to answer questions 1 and 2.

1. Which of the following could you do to the ball in the photograph above to cause a chemical change?
   A. cut in half   C. flatten
   B. paint   D. burn

2. Which of the following physical properties of the ball is size independent?
   A. density   C. volume
   B. mass   D. weight

3. Each of the following procedures results in the formation of bubbles. Which of these is a physical change?
   A. pouring an acid onto calcium carbonate
   B. dropping an antacid tablet into water
   C. heating water to its boiling point
   D. pouring vinegar onto baking soda

4. Which of the following occurs as you heat a liquid to its boiling point?
   A. condensation   C. vaporization
   B. melting   D. freezing

Test-Taking Tip

Essay Questions Spend a few minutes listing and organizing the main points that you plan to discuss. Make sure to do all of this work on your scratch paper, not on the answer sheet.

5. During an experiment, you find that you can dissolve 4.2 g of a substance in 250 mL of water at 25°C. How much of the substance would you predict that you could dissolve in 500 mL of water at the same temperature?
   A. 2.1 g   C. 6.3 g
   B. 4.2 g   D. 8.4 g

6. Which of the following is a chemical reaction?
   A. making ice cubes
   B. toasting bread
   C. slicing a carrot
   D. boiling water

7. When you make and eat scrambled eggs, many changes occur to the eggs. Which of the following best describes a chemical change?
   A. crack the eggs
   B. scramble the eggs
   C. cook the eggs
   D. chew the eggs

Use the table below to answer questions 8 and 9.

Physical Properties of Bromide	
Density	3.12 g/cm³
Boiling point	59°C
Melting point	−7°C

8. According to the table above, what is the mass of 4.34 cm³ of bromine?
   A. 0.719 g   C. 7.46 g
   B. 1.39 g   D. 13.5 g

9. At which of the following temperatures is bromine a solid?
   A. −10°C   C. 40°C
   B. 10°C   D. 80°C

Part 2 | Short Response/Grid In

Record your answers on the answer sheet provided by your teacher or on a sheet of paper.

10. A precipitate is one clue that a chemical change has occurred. What is a precipitate and when is it observed?

Use the photo below to answer questions 11 and 12.

11. The photograph above shows a rusted chain. Explain why rusting is a physical or a chemical change.

12. What are some physical properties of the rusty chain that you can see? What are some physical properties that you can't see?

13. You measure the density of a 12.3-g sample of limestone as 2.72 g/cm^3. What is the density of a 36.9 g sample?

14. A scientist measures the masses of two chemicals. He then combines the chemicals and measures their total mass. The total mass is less than the sum of each individual mass. Has this violated the law of conservation of mass? Explain what might have happened when the chemicals were combined.

15. A scientist measures 275 mL of water into a beaker. She then adds 51.0 g of lead into the beaker. After the addition of the lead, the volume of water in the beaker increases by 4.50 mL. What is the density of the lead?

Part 3 | Open Ended

Record your answers on a sheet of paper.

16. Suppose you have a gas in a closed container. Explain what would happen to the mass and density of the gas if you compressed it into half the volume.

17. Color change is an indication that a chemical change may have occurred. Mixing yellow and blue modeling clay makes green modeling clay. Is this a chemical reaction? Explain why or why not.

18. At a temperature of 40°C, you find that 40 g of ammonium chloride easily dissolves in 100 mL of water. When you stir 40 g of potassium chloride into a beaker containing 100 mL of water at 40°C, you find that some of the potassium chloride remains in the bottom of the beaker. Explain why this occurs and how to make to make the remaining potassium chloride dissolve.

Use the photo below to answer questions 19 and 20.

19. What would happen if you left the glass of cold water shown in the photograph above in the hot Sun for several hours? Describe how some physical properties of the water would change.

20. What properties of the water would not change? Explain why the density of the water would or would not change.

The Periodic Table

The BIG Idea

The periodic table provides information about all of the known elements.

SECTION 1
Introduction to the Periodic Table
Main Idea Elements are arranged in order of increasing atomic number on the periodic table.

SECTION 2
Representative Elements
Main Idea Representative elements within a group have similar properties.

SECTION 3
Transition Elements
Main Idea Transition elements are metals with a wide variety of uses.

Skyscrapers, Neon Lights, and the Periodic Table

Many cities have unique skylines. What is truly amazing is that everything in this photograph is made from 90 naturally occurring elements. In this chapter, you will learn more about the elements and the table that organizes them.

Science Journal Think of an element that you have heard about. Make a list of the properties you know and the properties you want to learn about.

Start-Up Activities

Make a Model of a Periodic Pattern

Every 29.5 days, the Moon begins to cycle through its phases from full moon to new moon and back again to full moon. Events that follow a predictable pattern are called periodic events. What other periodic events can you think of?

1. On a blank sheet of paper, make a grid with four squares across and four squares down.

2. Your teacher will give you 16 pieces of paper with different shapes and colors. Identify properties you can use to distinguish one piece of paper from another.

3. Place a piece of paper in each square on your grid. Arrange the pieces on the grid so that each column contains pieces that are similar.

4. Within each column, arrange the pieces to show a gradual change in their appearance.

5. **Think Critically** In your Science Journal, describe how the properties change in the rows across the grid and in the columns down the grid.

Preview this chapter's content and activities at
bookk.msscience.com

Periodic Table Make the following Foldable to help you classify the elements in the periodic table as metals, nonmetals, and metalloids.

| STEP 1 | Fold a vertical sheet of paper from side to side. Make the front edge about 1.25 cm shorter than the back edge. |

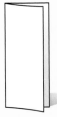

| STEP 2 | Turn lengthwise and **fold** into thirds. |

| STEP 3 | **Unfold and cut** only the top layer along both folds to make three tabs. **Label** each tab as shown. |

Metals | Metalloids | Nonmetals

Find Main Ideas As you read the chapter, write information about the three types of elements under the appropriate tabs. Use the information in your Foldable to explain how metalloids have properties between those of metals and nonmetals.

Get Ready to Read

Make Connections

① Learn It! Make connections between what you read and what you already know. Connections can be based on personal experiences (text-to-self), what you have read before (text-to-text), or events in other places (text-to-world).

As you read, ask connecting questions. Are you reminded of a personal experience? Have you read about the topic before? Did you think of a person, a place, or an event in another part of the world?

② Practice It! Read the excerpt below and make connections to your own knowledge and experience.

Text-to-self:
What metals do you use every day?

Text-to-text:
What have you read about melting points before?

Text-to-world:
Have you heard about mercury in the news or seen a mercury thermometer?

> If you look at the periodic table, you will notice it is color coded. The colors represent elements that are metals, nonmetals, or metalloids. With the exception of mercury, all the metals are solids, most with high melting points. A metal is an element that has luster, is a good conductor of heat and electricity, is malleable, and is ductile.
>
> —*from page 100*

③ Apply It! As you read this chapter, choose five words or phrases that make a connection to something you already know.

Target Your Reading

Reading Tip

Make connections with memorable events, places, or people in your life. The better the connection, the more likely you will to be remember it.

Use this to focus on the main ideas as you read the chapter.

1 **Before you read** the chapter, respond to the statements below on your worksheet or on a numbered sheet of paper.

- Write an **A** if you **agree** with the statement.
- Write a **D** if you **disagree** with the statement.

2 **After you read** the chapter, look back to this page to see if you've changed your mind about any of the statements.

- If any of your answers changed, explain why.
- Change any false statements into true statements.
- Use your revised statements as a study guide.

Science Online
Print out a worksheet of this page at bookk.msscience.com

Before You Read A or D		Statement	After You Read A or D
	1	Scientists have discovered all the elements that could possibly exist.	
	2	The elements are arranged on the periodic table according to their atomic numbers and mass numbers.	
	3	Elements in a group have similar properties.	
	4	Metals are located on the right side of the periodic table.	
	5	When a new element is discovered, the IUPAC selects a name.	
	6	Only metals conduct electricity.	
	7	Noble gases rarely combine with other elements.	
	8	The transition elements contain metals, nonmetals, and metalloids.	
	9	Some elements are created in a lab.	

Introduction to the Periodic Table

as you read

What You'll Learn

- **Describe** the history of the periodic table.
- **Interpret** an element key.
- **Explain** how the periodic table is organized.

Why It's Important

The periodic table makes it easier for you to find information that you need about the elements.

🔍 **Review Vocabulary**

element: a substance that cannot be broken down into simpler substances

New Vocabulary
- period
- group
- representative element
- transition element
- metal
- nonmetal
- metalloid

Development of the Periodic Table

Early civilizations were familiar with a few of the substances now called elements. They made coins and jewelry from gold and silver. They also made tools and weapons from copper, tin, and iron. In the nineteenth century, chemists began to search for new elements. By 1830, they had isolated and named 55 different elements. The list continues to grow today.

Mendeleev's Table of Elements A Russian chemist, Dmitri Mendeleev (men duh LAY uhf), published the first version of his periodic table in the *Journal of the Russian Chemical Society* in 1869. His table is shown in **Figure 1.** When Mendeleev arranged the elements in order of increasing atomic mass, he began to see a pattern. Elements with similar properties fell into groups on the table. At that time, not all the elements were known. To make his table work, Mendeleev had to leave three gaps for missing elements. Based on the groupings in his table, he predicted the properties for the missing elements. Mendeleev's predictions spurred other chemists to look for the missing elements. Within 15 years, all three elements—gallium, scandium, and germanium—were discovered.

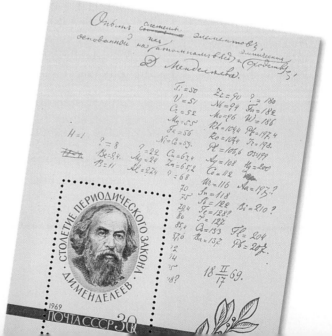

Figure 1 Mendeleev published his first periodic table in 1869. This postage stamp, with his table and photo, was issued in 1969 to commemorate the event. Notice the question marks that he used to mark his prediction of yet-undiscovered elements.

Moseley's Contribution Although Mendeleev's table correctly organized most of the elements, a few elements seemed out of place. In the early twentieth century, the English physicist Henry Moseley, before age 27, realized that Mendeleev's table could be improved by arranging the elements according to atomic number rather than atomic mass. Moseley revised the periodic table by arranging the elements in order of increasing number of protons in the nucleus. With Moseley's table, it was clear how many elements still were undiscovered.

Today's Periodic Table

In the modern periodic table on the next page, the elements still are organized by increasing atomic number. The rows or periods are labeled 1–7. A **period** is a row of elements in the periodic table whose properties change gradually and predictably. The periodic table has 18 columns of elements. Each column contains a group, or family, of elements. A **group** contains elements that have similar physical or chemical properties.

Zones on the Periodic Table The periodic table can be divided into sections, as you can see in **Figure 2.** One section consists of the first two groups, Groups 1 and 2, and the elements in Groups 13–18. These eight groups are the **representative elements.** They include metals, metalloids, and nonmetals. The elements in Groups 3–12 are **transition elements.** They are all metals. Some transition elements, called the inner transition elements, are placed below the main table. These elements are called the lanthanide and actinide series because one series follows the element lanthanum, element 57, and the other series follows actinium, element 89.

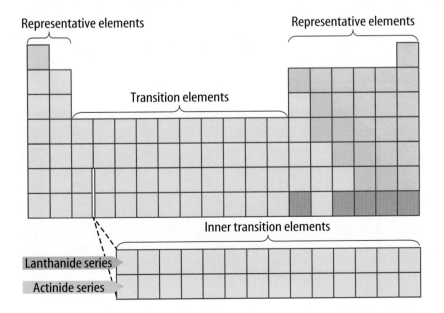

Representative elements Representative elements
Transition elements
Inner transition elements
Lanthanide series
Actinide series

Figure 2 The periodic table is divided into sections. Traditionally, the lanthanides and actinides are placed below the table so that the table will not be as wide. These elements have similar properties. **Identify** *the section of the periodic table that contains only metal.*

PERIODIC TABLE OF THE ELEMENTS

Columns of elements are called groups. Elements in the same group have similar chemical properties.

Gas
Liquid
Solid
Synthetic

Element —— Hydrogen
Atomic number —— 1
Symbol —— **H**
Atomic mass —— 1.008
State of matter

The first three symbols tell you the state of matter of the element at room temperature. The fourth symbol identifies elements that are not present in significant amounts on Earth. Useful amounts are made synthetically.

1	2	3	4	5	6	7	8	9
1 Hydrogen 1 **H** 1.008								
2 Lithium 3 **Li** 6.941	Beryllium 4 **Be** 9.012							
3 Sodium 11 **Na** 22.990	Magnesium 12 **Mg** 24.305							
4 Potassium 19 **K** 39.098	Calcium 20 **Ca** 40.078	Scandium 21 **Sc** 44.956	Titanium 22 **Ti** 47.867	Vanadium 23 **V** 50.942	Chromium 24 **Cr** 51.996	Manganese 25 **Mn** 54.938	Iron 26 **Fe** 55.845	Cobalt 27 **Co** 58.933
5 Rubidium 37 **Rb** 85.468	Strontium 38 **Sr** 87.62	Yttrium 39 **Y** 88.906	Zirconium 40 **Zr** 91.224	Niobium 41 **Nb** 92.906	Molybdenum 42 **Mo** 95.94	Technetium 43 **Tc** (98)	Ruthenium 44 **Ru** 101.07	Rhodium 45 **Rh** 102.906
6 Cesium 55 **Cs** 132.905	Barium 56 **Ba** 137.327	Lanthanum 57 **La** 138.906	Hafnium 72 **Hf** 178.49	Tantalum 73 **Ta** 180.948	Tungsten 74 **W** 183.84	Rhenium 75 **Re** 186.207	Osmium 76 **Os** 190.23	Iridium 77 **Ir** 192.217
7 Francium 87 **Fr** (223)	Radium 88 **Ra** (226)	Actinium 89 **Ac** (227)	Rutherfordium 104 **Rf** (261)	Dubnium 105 **Db** (262)	Seaborgium 106 **Sg** (266)	Bohrium 107 **Bh** (264)	Hassium 108 **Hs** (277)	Meitnerium 109 **Mt** (268)

The number in parentheses is the mass number of the longest-lived isotope for that element.

Rows of elements are called periods. Atomic number increases across a period.

The arrow shows where these elements would fit into the periodic table. They are moved to the bottom of the table to save space.

Lanthanide series

Cerium 58 **Ce** 140.116	Praseodymium 59 **Pr** 140.908	Neodymium 60 **Nd** 144.24	Promethium 61 **Pm** (145)	Samarium 62 **Sm** 150.36

Actinide series

Thorium 90 **Th** 232.038	Protactinium 91 **Pa** 231.036	Uranium 92 **U** 238.029	Neptunium 93 **Np** (237)	Plutonium 94 **Pu** (244)

Metal

Metalloid

Nonmetal

The color of an element's block tells you if the element is a metal, nonmetal, or metalloid.

Science Online

Visit bookk.msscience.com for updates to the periodic table.

13	14	15	16	17	18
					Helium 2 **He** 4.003
Boron 5 **B** 10.811	Carbon 6 **C** 12.011	Nitrogen 7 **N** 14.007	Oxygen 8 **O** 15.999	Fluorine 9 **F** 18.998	Neon 10 **Ne** 20.180
Aluminum 13 **Al** 26.982	Silicon 14 **Si** 28.086	Phosphorus 15 **P** 30.974	Sulfur 16 **S** 32.065	Chlorine 17 **Cl** 35.453	Argon 18 **Ar** 39.948

10	11	12	13	14	15	16	17	18
Nickel 28 **Ni** 58.693	Copper 29 **Cu** 63.546	Zinc 30 **Zn** 65.409	Gallium 31 **Ga** 69.723	Germanium 32 **Ge** 72.64	Arsenic 33 **As** 74.922	Selenium 34 **Se** 78.96	Bromine 35 **Br** 79.904	Krypton 36 **Kr** 83.798
Palladium 46 **Pd** 106.42	Silver 47 **Ag** 107.868	Cadmium 48 **Cd** 112.411	Indium 49 **In** 114.818	Tin 50 **Sn** 118.710	Antimony 51 **Sb** 121.760	Tellurium 52 **Te** 127.60	Iodine 53 **I** 126.904	Xenon 54 **Xe** 131.293
Platinum 78 **Pt** 195.078	Gold 79 **Au** 196.967	Mercury 80 **Hg** 200.59	Thallium 81 **Tl** 204.383	Lead 82 **Pb** 207.2	Bismuth 83 **Bi** 208.980	Polonium 84 **Po** (209)	Astatine 85 **At** (210)	Radon 86 **Rn** (222)
Darmstadtium 110 **Ds** (281)	Roentgenium 111 **Rg** (272)	Ununbium * 112 **Uub** (285)		Ununquadium * 114 **Uuq** (289)				

✳ The names and symbols for elements 112 and 114 are temporary. Final names will be selected when the elements' discoveries are verified.

Europium 63 **Eu** 151.964	Gadolinium 64 **Gd** 157.25	Terbium 65 **Tb** 158.925	Dysprosium 66 **Dy** 162.500	Holmium 67 **Ho** 164.930	Erbium 68 **Er** 167.259	Thulium 69 **Tm** 168.934	Ytterbium 70 **Yb** 173.04	Lutetium 71 **Lu** 174.967
Americium 95 **Am** (243)	Curium 96 **Cm** (247)	Berkelium 97 **Bk** (247)	Californium 98 **Cf** (251)	Einsteinium 99 **Es** (252)	Fermium 100 **Fm** (257)	Mendelevium 101 **Md** (258)	Nobelium 102 **No** (259)	Lawrencium 103 **Lr** (262)

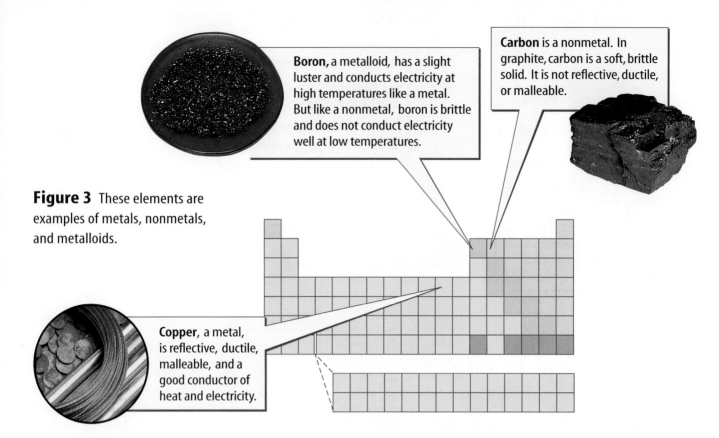

Boron, a metalloid, has a slight luster and conducts electricity at high temperatures like a metal. But like a nonmetal, boron is brittle and does not conduct electricity well at low temperatures.

Carbon is a nonmetal. In graphite, carbon is a soft, brittle solid. It is not reflective, ductile, or malleable.

Figure 3 These elements are examples of metals, nonmetals, and metalloids.

Copper, a metal, is reflective, ductile, malleable, and a good conductor of heat and electricity.

Science Online

Topic: Elements

Visit bookk.msscience.com for Web links to information about how the periodic table was developed.

Activity Select an element and write about how, when, and by whom it was discovered.

Metals If you look at the periodic table, you will notice it is color coded. The colors represent elements that are metals, nonmetals, or metalloids. Examples of a metal, a nonmetal, and a metalloid are illustrated in **Figure 3.** With the exception of mercury, all the metals are solids, most with high melting points. A **metal** is an element that has luster, is a good conductor of heat and electricity, is malleable, and is ductile. The ability to reflect light is a property of metals called luster. Many metals can be pressed or pounded into thin sheets or shaped into objects because they are malleable (MAL yuh bul). Metals are also ductile (DUK tul), which means that they can be drawn out into wires. Can you think of any items that are made of metals?

Nonmetals and Metalloids **Nonmetals** are usually gases or brittle solids at room temperature and poor conductors of heat and electricity. There are only 17 nonmetals, but they include many elements that are essential for life—carbon, sulfur, nitrogen, oxygen, phosphorus, and iodine.

The elements between metals and nonmetals on the periodic table are called metalloids (ME tuh loydz). As you might expect from the name, a **metalloid** is an element that shares some properties with metals and some with nonmetals. These elements also are called semimetals.

✓ **Reading Check** *How many elements are nonmetals?*

The Element Keys Each element is represented on the periodic table by a box called the element key. An enlarged key for hydrogen is shown in **Figure 4.** An element key shows you the name of the element, its atomic number, its symbol, and its average atomic mass. Element keys for elements that occur naturally on Earth include a logo that tells whether the element is a solid, a liquid, or a gas at room temperature. All the gases except hydrogen are on the right side of the table. They are marked with a balloon logo. Most of the other elements are solids at room temperature and are marked with a cube. Two elements on the periodic table are liquids at room temperature. Their logo is a drop. Elements that do not occur naturally on Earth are marked with a bull's-eye logo. These are synthetic elements.

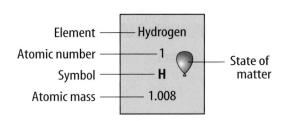

Element —— Hydrogen
Atomic number —— 1
Symbol —— H
Atomic mass —— 1.008
State of matter

Figure 4 As you can see from the element key, a lot of information about an element is given on the periodic table.
Identify *the two elements that are liquids at room temperature.*

Applying Science

What does *periodic* mean in the periodic table?

Elements often combine with oxygen to form oxides and chlorine to form chlorides. For example, two hydrogen atoms combine with one oxygen atom to form oxide, H_2O or water. One sodium atom combines with one chlorine atom to form sodium chloride, NaCl or table salt. The location of an element on the periodic table is an indication of how it combines with other elements.

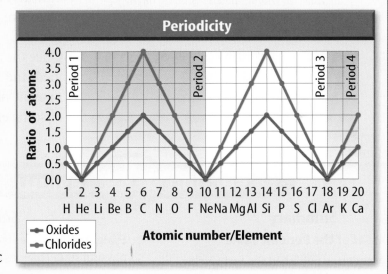

Periodicity

Ratio of atoms

Period 1 Period 2 Period 3 Period 4

4.0
3.5
3.0
2.5
2.0
1.5
1.0
0.5
0.0

1 2 3 4 5 6 7 8 9 10 11 12 13 14 15 16 17 18 19 20
H He Li Be B C N O F Ne Na Mg Al Si P S Cl Ar K Ca

Oxides
Chlorides

Atomic number/Element

Identifying the Problem

The graph shows the number of oxygen atoms (red) and chlorine atoms (green) that will combine with the first 20 elements. What pattern do you see?

Solving the Problem

1. Find all of the elements in Group 1 on the graph. Do the same with the elements in Groups 14 and 18. What do you notice about their positions on the graph?
2. This relationship demonstrates one of the properties of a group of elements. Follow the elements in order on the periodic table and on the graph. Write a statement using the word *periodic* that describes what occurs with the elements and their properties.

Table 1 Chemical Symbols and Their Origins

Name	Symbol	Origin of Name
Mendelevium	Md	For Dimitri Mendeleev
Lead	Pb	The Latin name for lead is *plumbum*.
Thorium	Th	The Norse god of thunder is Thor.
Polonium	Po	For Poland, where Marie Curie, a famous scientist, was born
Hydrogen	H	From Greek words meaning "water former"
Mercury	Hg	*Hydrargyrum* means "liquid silver" in Greek.
Gold	Au	*Aurum* means "shining dawn" in Latin.
Unununium	Uuu	Named using the IUPAC naming system

Symbols for the Elements The symbols for the elements are either one- or two-letter abbreviations, often based on the element name. For example, V is the symbol for vanadium, and Sc is the symbol for scandium. Sometimes the symbols don't match the names. Examples are Ag for silver and Na for sodium. In those cases, the symbol might come from Greek or Latin names for the elements. Some elements are named for scientists such as Lise Meitner (meitnerium, Mt). Some are named for geographic locations such as France (francium, Fr).

Newly synthesized elements are given a temporary name and 3-letter symbol that is related to the element's atomic number. The International Union of Pure and Applied Chemistry (IUPAC) adopted this system in 1978. Once the discovery of the element is verified, the discoverers can choose a permanent name. **Table 1** shows the origin of some element names and symbols.

section 1 review

Summary

Development of the Periodic Table

- Dmitri Mendeleev published the first version of the periodic table in 1869.
- Mendeleev left three gaps on the periodic table for missing elements.
- Moseley arranged Mendeleev's table according to atomic number, not by atomic mass.

Today's Periodic Table

- The periodic table is divided into sections.
- A period is a row of elements whose properties change gradually and predictably.
- Groups 1 and 2 along with Groups 13–18 are called representative elements.
- Groups 3–12 are called transition elements.

Self Check

1. **Evaluate** the elements in period 4 to show how the physical state changes as the atomic number increases.
2. **Describe** where the metals, nonmetals, and metalloids are located in the periodic table.
3. **Classify** each of the following elements as metal, nonmetal, or metalloid: Fe, Li, B, Cl, Si, Na, and Ni.
4. **List** what an element key contains.
5. **Think Critically** How would the modern periodic table be different if elements were arranged by average atomic mass instead of by atomic number?

Applying Math

6. **Solve One-Step Equations** What is the difference in atomic mass of iodine and magnesium?

section
2

Representative Elements

Groups 1 and 2

Groups 1 and 2 are always found in nature combined with other elements. They're called active metals because of their readiness to form new substances with other elements. They are all metals except hydrogen, the first element in Group 1. Although hydrogen is placed in Group 1, it shares properties with the elements in Group 1 and Group 17.

Alkali Metals The Group 1 elements have a specific family name—**alkali metals.** All the alkali metals are silvery solids with low densities and low melting points. These elements increase in their reactivity, or tendency to combine with other substances, as you move from top to bottom on the periodic table. Some uses of the alkali metals are shown in **Figure 5.**

Alkali metals are found in many items. Lithium batteries are used in cameras. Sodium chloride is common table salt. Sodium and potassium, dietary requirements, are found in small quantities in potatoes and bananas.

as you read

***What* You'll Learn**
- **Recognize** the properties of representative elements.
- **Identify** uses for the representative elements.
- **Classify** elements into groups based on similar properties.

***Why* It's Important**
Many representative elements play key roles in your body, your environment, and in the things you use every day.

Review Vocabulary
atomic number: the number of protons in the nucleus of a given element

New Vocabulary
- alkali metal
- alkaline earth metal
- semiconductor
- halogen
- noble gas

Figure 5 These items contain alkali metals.

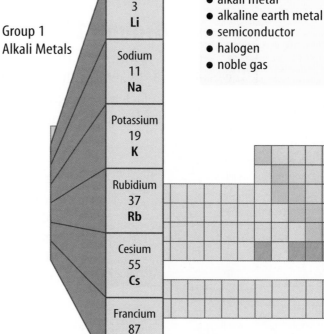

Group 1
Alkali Metals

| Lithium 3 Li |
| Sodium 11 Na |
| Potassium 19 K |
| Rubidium 37 Rb |
| Cesium 55 Cs |
| Francium 87 Fr |

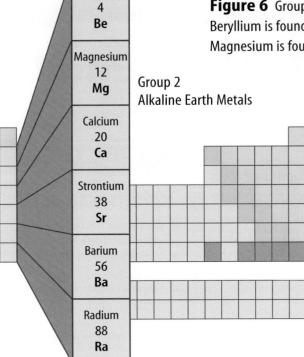

Beryllium
4
Be

Magnesium
12
Mg

Group 2
Alkaline Earth Metals

Calcium
20
Ca

Strontium
38
Sr

Barium
56
Ba

Radium
88
Ra

Figure 6 Group 2 elements are found in many things. Beryllium is found in the gems emerald and aquamarine. Magnesium is found in the chlorophyll of green plants.

Alkaline Earth Metals Next door to the alkali metals' family are their Group 2 neighbors, the **alkaline earth metals.** Each alkaline earth metal is denser and harder and has a higher melting point than the alkali metal in the same period. Alkaline earth metals are reactive, but not as reactive as the alkali metals. Some uses of the alkaline earth elements are shown in **Figure 6.**

 *What are the names of the elements that are alkaline earth metals?*

Groups 13 through 18

Notice on the periodic table that the elements in Groups 13–18 are not all solid metals like the elements of Groups 1 and 2. In fact, a single group can contain metals, nonmetals, and metalloids and have members that are solids, liquids, and gases.

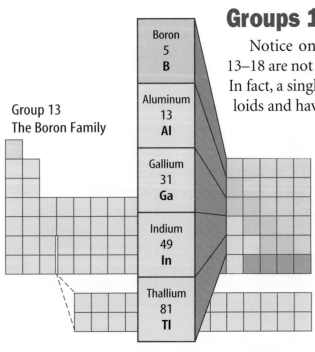

Group 13
The Boron Family

Boron
5
B

Aluminum
13
Al

Gallium
31
Ga

Indium
49
In

Thallium
81
Tl

Group 13—The Boron Family The elements in Group 13 are all metals except boron, which is a brittle, black metalloid. This family of elements is used to make a variety of products. Cookware made with boron can be moved directly from the refrigerator into the oven without cracking. Aluminum is used to make soft-drink cans, cookware, siding for homes, and baseball bats. Gallium is a solid metal, but its melting point is so low that it will melt in your hand. It is used to make computer chips.

Group 14—The Carbon Group If you look at Group 14, you can see that carbon is a nonmetal, silicon and germanium are metalloids, and tin and lead are metals. The nonmetal carbon exists as an element in several forms. You're familiar with two of them—diamond and graphite. Carbon also is found in all living things. Carbon is followed by the metalloid silicon, an abundant element contained in sand. Sand contains ground-up particles of minerals such as quartz, which is composed of silicon and oxygen. Glass is an important product made from sand.

Silicon and its Group 14 neighbor, germanium, are metalloids. They are used in electronics as semiconductors. A **semiconductor** doesn't conduct electricity as well as a metal, but does conduct electricity better than a nonmetal. Silicon and small amounts of other elements are used for computer chips as shown in **Figure 7.**

Tin and lead are the two heaviest elements in Group 14. Lead is used in the apron, shown in **Figure 7,** to protect your torso during dental X rays. It also is used in car batteries, low-melting alloys, protective shielding around nuclear reactors, particle accelerators, X-ray equipment, and containers used for storing and transporting radioactive materials. Tin is used in pewter, toothpaste, and the coating on steel cans used for food.

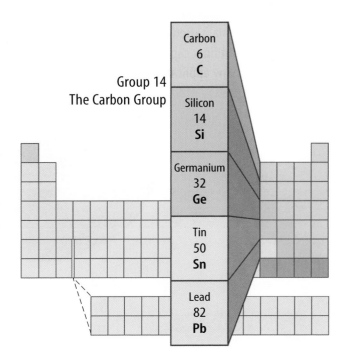

Group 14
The Carbon Group

| Carbon 6 **C** |
| Silicon 14 **Si** |
| Germanium 32 **Ge** |
| Tin 50 **Sn** |
| Lead 82 **Pb** |

Figure 7 Members of Group 14 include one nonmetal, two metalloids, and two metals.

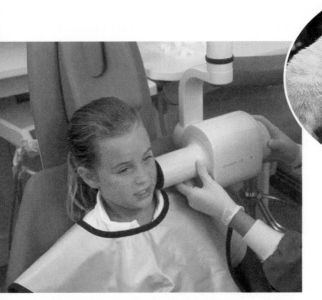

Lead is used to shield your body from unwanted X-ray exposure.

All living things contain carbon compounds.

Silicon crystals are used to make computer chips.

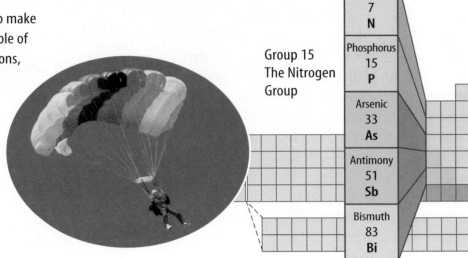

Figure 8 Ammonia is used to make nylon, a tough, light fiber capable of replacing silk in many applications, including parachutes.

Group 15
The Nitrogen
Group

Nitrogen 7 N
Phosphorus 15 P
Arsenic 33 As
Antimony 51 Sb
Bismuth 83 Bi

Farmers Each year farmers test their soil to determine the level of nutrients, the matter needed for plants to grow. The results of the test help the farmer decide how much nitrogen, phosphorus, and potassium to add to the fields. The additional nutrients increase the chance of having a successful crop.

Figure 9 Nitrogen and phosphorus are required for healthy green plants. This synthetic fertilizer label shows the nitrogen and phosphorous compounds that provide these.

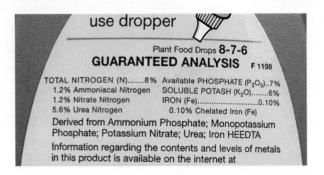

use dropper

Plant Food Drops **8-7-6**
GUARANTEED ANALYSIS F 1198

TOTAL NITROGEN (N)......8% Available PHOSPHATE (P_2O_5)..7%
 1.2% Ammoniacal Nitrogen SOLUBLE POTASH (K_2O)........6%
 1.2% Nitrate Nitrogen IRON (Fe)...........................0.10%
 5.6% Urea Nitrogen 0.10% Chelated Iron (Fe)
Derived from Ammonium Phosphate; Monopotassium Phosphate; Potassium Nitrate; Urea; Iron HEEDTA
Information regarding the contents and levels of metals in this product is available on the internet at

Group 15—The Nitrogen Group At the top of Group 15 are the two nonmetals—nitrogen and phosphorus. Nitrogen and phosphorus are required by living things and are used to manufacture various items. These elements also are parts of the biological materials that store genetic information and energy in living organisms. Although almost 80 percent of the air you breathe is nitrogen, you can't get the nitrogen your body needs by breathing nitrogen gas. Bacteria in the soil must first change nitrogen gas into substances that can be absorbed through the roots of plants. Then, by eating the plants, nitrogen becomes available to your body.

Reading Check *Can your body obtain nitrogen by breathing air? Explain.*

Ammonia is a gas that contains nitrogen and hydrogen. When ammonia is dissolved in water, it can be used as a cleaner and disinfectant. Liquid ammonia is sometimes applied directly to soil as a fertilizer. Ammonia also can be converted into solid fertilizers. It also is used to freeze-dry food and as a refrigerant. Ammonia also is used to make nylon for parachutes, as shown in **Figure 8.**

The element phosphorus comes in two forms—white and red. White phosphorus is so active it can't be exposed to oxygen in the air or it will burst into flames. The heads of matches contain the less active red phosphorus, which ignites from the heat produced by friction when the match is struck. Phosphorous compounds are essential ingredients for healthy teeth and bones. Plants also need phosphorus, so it is one of the nutrients in most fertilizers. The fertilizer label in **Figure 9** shows the compounds of nitrogen and phosphorus that are used to give plants a synthetic supply of these elements.

Group 16—The Oxygen Family

The first two members of Group 16, oxygen and sulfur, are essential for life. The heavier members of the group, tellurium and polonium, are both metalloids.

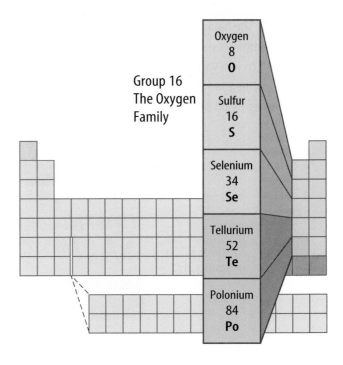

About 20 percent of Earth's atmosphere is the oxygen you breathe. Your body needs oxygen to release the energy from the foods you eat. Oxygen is abundant in Earth's rocks and minerals because it readily combines with other elements. Oxygen also is required for combustion to occur. Foam is used in fire fighting to keep oxygen away from the burning item, as shown in **Figure 10.** Ozone, a less common form of oxygen, is formed in the upper atmosphere through the action of electricity during thunderstorms. The presence of ozone is important because it shields living organisms from some harmful radiation from the Sun.

Sulfur is a solid, yellow nonmetal. Large amounts of sulfur are used to manufacture sulfuric acid, one of the most commonly used chemicals in the world. Sulfuric acid is a combination of sulfur, hydrogen, and oxygen. It is used in the manufacture of paints, fertilizers, detergents, synthetic fibers, and rubber.

Selenium conducts electricity when exposed to light, so it is used in solar cells, light meters, and photographic materials. Its most important use is as the light-sensitive component in photocopy machines. Traces of selenium are also necessary for good health.

Poison Buildup Arsenic disrupts the normal function of an organism by disrupting cellular metabolism. Because arsenic builds up in hair, forensic scientists can test hair samples to confirm or disprove a case of arsenic poisoning. Tests of Napoleon's hair suggest that he was poisoned with arsenic. Use reference books to find out who Napoleon I was and why someone might have wanted to poison him.

Figure 10 The foam used in aircraft fires forms a film of water over the burning fuel which suffocates the fire.

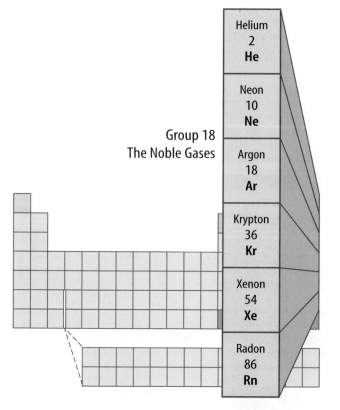

Figure 11 The halogens are a group of elements that are important to us in a variety of ways. Chlorine is added to drinking water to kill bacteria.

Iodine is needed by many systems in your body.

Group 17
The Halogen
Group

| Fluorine |
| 9 |
| **F** |

| Chlorine |
| 17 |
| **Cl** |

| Bromine |
| 35 |
| **Br** |

| Iodine |
| 53 |
| **I** |

| Astatine |
| 85 |
| **At** |

Group 17—The Halogen Group All the elements in Group 17 are nonmetals except for astatine, which is a radioactive metalloid. These elements are called **halogens,** which means "salt-former." Table salt, sodium chloride, is a substance made from sodium and chlorine. All of the halogens form similar salts with sodium and with the other alkali metals.

The halogen fluorine is the most reactive of the halogens in combining with other elements. Chlorine is less reactive than fluorine, and bromine is less reactive than chlorine. Iodine is the least reactive of the four nonmetals. **Figure 11** shows some uses of halogens.

✓ Reading Check *What do halogens form with the alkali metals?*

Group 18—The Noble Gases The Group 18 elements are called the **noble gases.** This is because they rarely combine with other elements and are found only as uncombined elements in nature. Their reactivity is very low.

Helium is less dense than air, so it's great for all kinds of balloons, from party balloons to blimps that carry television cameras high above sporting events. Helium balloons, such as the one in **Figure 12,** lift instruments into the upper atmosphere to measure atmospheric conditions. Even though hydrogen is lighter than helium, helium is preferred for these purposes because helium will not burn.

| Helium |
| 2 |
| **He** |

| Neon |
| 10 |
| **Ne** |

| Argon |
| 18 |
| **Ar** |

| Krypton |
| 36 |
| **Kr** |

| Xenon |
| 54 |
| **Xe** |

| Radon |
| 86 |
| **Rn** |

Group 18
The Noble Gases

Uses for the Noble Gases The "neon" lights you see in advertising signs, like the one in **Figure 12,** can contain any of the noble gases, not just neon. Electricity is passed through the glass tubes that make up the sign. These tubes contain the noble gas, and the electricity causes the gas to glow. Each noble gas produces a unique color. Helium glows yellow, neon glows red-orange, and argon produces a bluish-violet color.

Argon, the most abundant of the noble gases on Earth, was first found in 1894. Krypton is used with nitrogen in ordinary lightbulbs because these gases keep the glowing filament from burning out. When a mixture of argon, krypton, and xenon is used, a bulb can last longer than bulbs that do not contain this mixture. Krypton lights are used to illuminate landing strips at airports, and xenon is used in strobe lights and was once used in photographic flash cubes.

At the bottom of the group is radon, a radioactive gas produced naturally as uranium decays in rocks and soil. If radon seeps into a home, the gas can be harmful because it continues to emit radiation. When people breathe the gas over a period of time, it can cause lung cancer.

Figure 12 Noble gases are used in many applications. Scientists use helium balloons to measure atmospheric conditions.

Each noble gas glows a different color when an electric current is passed through it.

✔ **Reading Check** *Why are noble gases used in lights?*

section 2 review

Summary

Groups 1 and 2

- Groups 1 and 2 elements are always combined with other elements.
- The elements in Groups 1 and 2 are all metals except for hydrogen.
- Alkaline earth metals are not as active as the alkali metals.

Groups 13–18

- With Groups 13–18, a single group can contain metals, nonmetals, and metalloids.
- Nitrogen and phosphorus are required by living things.
- The halogen group will form salts with alkali metals.

Self Check

1. **Compare and contrast** the elements in Group 1 and the elements in Group 17.
2. **Describe** two uses for a member of each representative group.
3. **Identify** the group of elements that does not readily combine with other elements.
4. **Think Critically** Francium is a rare radioactive alkali metal at the bottom of Group 1. Its properties have not been studied carefully. Would you predict that francium would combine with water more or less readily than cesium?

Applying Skills

5. **Predict** how readily astatine would form a salt compared to the other elements in Group 17. Is there a trend for reactivity in this group?

Transition Elements

What You'll Learn

- **Identify** properties of some transition elements.
- **Distinguish** lanthanides from actinides.

Why It's Important

Transition elements provide the materials for many things including electricity in your home and steel for construction.

⚙ Review Vocabulary

mass number: the sum of neutrons and protons in the nucleus of an atom

New Vocabulary

- catalyst
- lanthanide
- actinide
- synthetic element

Figure 13 These buildings and bridges have steel in their structure. **Explain** *why you think steel is used in their construction.*

The Metals in the Middle

Groups 3–12 are called the transition elements and all of them are metals. Across any period from Group 3 through 12, the properties of the elements change less noticeably than they do across a period of representative elements.

Most transition elements are found combined with other elements in ores. A few transition elements such as gold and silver are found as pure elements.

The Iron Triad Three elements in period 4—iron, cobalt, and nickel—have such similar properties that they are known as the iron triad. These elements, among others, have magnetic properties. Industrial magnets are made from an alloy of nickel, cobalt, and aluminum. Nickel is used in batteries along with cadmium. Iron is a necessary part of hemoglobin, the substance that transports oxygen in the blood.

Iron also is mixed with other metals and with carbon to create a variety of steels with different properties. Structures such as bridges and skyscrapers, shown in **Figure 13,** depend upon steel for their strength.

✔ **Reading Check** *Which metals make up the iron triad?*

The Iron Triad

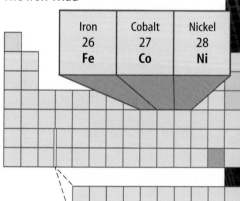

| | Iron 26 **Fe** | Cobalt 27 **Co** | Nickel 28 **Ni** | |

Uses of Transition Elements Most transition metals have higher melting points than the representative elements. The filaments of lightbulbs, like the one in **Figure 14,** are made of tungsten, element 74. Tungsten has the highest melting point of any metal (3,410°C) and will not melt when a current passes through it.

Mercury, which has the lowest melting point of any metal (−39°C), is used in thermometers and in barometers. Mercury is the only metal that is a liquid at room temperatures. Like many of the heavy metals, mercury is poisonous to living beings. Therefore, mercury must be handled with care.

Chromium's name comes from the Greek word for color, *chroma,* and the element lives up to its name. Two substances containing chromium are shown in **Figure 15.** Many other transition elements combine to form substances with equally brilliant colors.

Ruthenium, rhodium, palladium, osmium, iridium, and platinum are sometimes called the platinum group because they have similar properties. They do not combine easily with other elements. As a result, they can be used as catalysts. A **catalyst** is a substance that can make something happen faster but is not changed itself. Other transition elements, such as nickel, zinc, and cobalt, can be used as catalysts. As catalysts, the transition elements are used to produce electronic and consumer goods, plastics, and medicines.

Figure 14 The transition metal tungsten is used in lightbulbs because of its high melting point.

Figure 15 Transition metals are used in a variety of products.

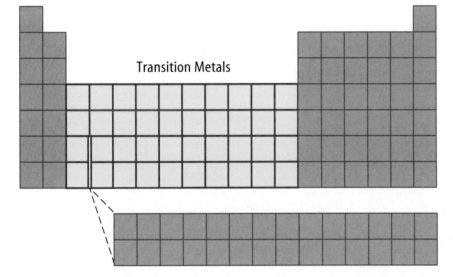

Transition Metals

Inner Transition Elements

There are two series of inner transition elements. The first series, from cerium to lutetium, is called the **lanthanides.** The lanthanides also are called the rare earths because at one time they were thought to be scarce. The lanthanides are usually found combined with oxygen in Earth's crust. The second series of elements, from thorium to lawrencium, is called the **actinides.**

✔ **Reading Check** *What other name is used to refer to the lanthanides?*

The Lanthanides The lanthanides are soft metals that can be cut with a knife. The elements are so similar that they are hard to separate when they occur in the same ore, which they often do. Despite the name rare earth, the lanthanides are not as rare as originally thought. Earth's crust contains more cerium than lead. Cerium makes up 50 percent of an alloy called misch (MIHSH) metal. Flints in lighters, like the one in **Figure 16,** are made from misch metal. The other ingredients in flint are lanthanum, neodymium, and iron.

The Actinides All the actinides are radioactive. The nuclei of atoms of radioactive elements are unstable and decay to form other elements. Thorium, protactinium, and uranium are the only actinides that now are found naturally on Earth. Uranium is found in Earth's crust because its half-life is long—4.5 billion years. All other actinides are synthetic elements. **Synthetic elements** are made in laboratories and nuclear reactors. **Figure 17** shows how synthetic elements are made. The synthetic elements have many uses. Plutonium is used as a fuel in nuclear power plants. Americium is used in some home smoke detectors. Californium-252 is used to kill cancer cells.

✔ **Reading Check** *What property do all actinides share?*

Figure 16 The flint in this lighter is called misch metal, which is about 50% cerium, 25% lanthanum, 15% neodymium, and 10% other rare earth metals and iron.

Inner Transition Elements

Lanthanide Series	58 Ce	59 Pr	60 Nd	61 Pm	62 Sm	63 Eu	64 Gd	65 Tb	66 Dy	67 Ho	68 Er	69 Tm	70 Yb	71 Lu
Actinide Series	90 Th	91 Pa	92 U	93 Np	94 Pu	95 Am	96 Cm	97 Bk	98 Cf	99 Es	100 Fm	101 Md	102 No	103 Lr

Figure 17

No element heavier than uranium, with 92 protons and 146 neutrons, is typically found in nature. But by using a device called a particle accelerator, scientists can make synthetic elements with atomic numbers greater than that of uranium. Within the accelerator, atomic nuclei are made to collide at high speeds in the hope that some will fuse together to form new, heavier elements. These "heavy" synthetic elements are radioactive isotopes, some of which are so unstable that they survive only a fraction of a second before emitting radioactive particles and decaying into other, lighter elements.

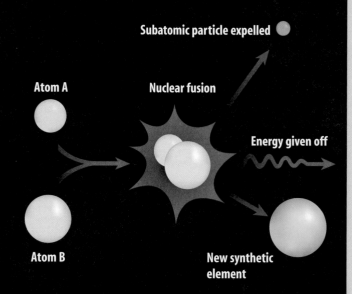

▲ When atoms collide in an accelerator, their nuclei may undergo a fusion reaction to form a new—and often short-lived—synthetic element. Energy and one or more subatomic particles typically are given off in the process.

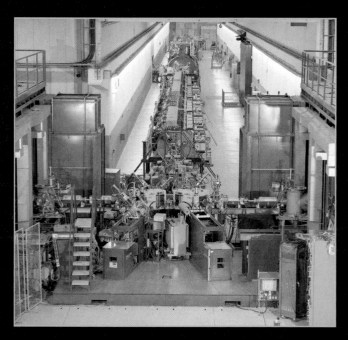

▲ Inside the airless vacuum chamber of a particle accelerator, such as this one in Hesse, Germany, streams of atoms move at incredibly high speeds.

▶ Recently, the IUPAC (International Union of Pure and Applied Chemistry) General Assembly confirmed the official name and symbol of element 110. Element 110 was previously known as Ununnilium and its symbol was Uun. The new name is darmstadtium and its symbol is Ds. Element 111 is expected to receive its official name and symbol in the near future.

Dentistry and Dental Materials

Dentists have been using amalgam for over 150 years to fill cavities in decayed teeth. Amalgam, a mixture of silver, copper, tin, and mercury, is the familiar "silver filling." Because amalgam contains mercury, some people are concerned that the use of this type of filling may unnecessarily expose a person to mercury vapor. Today dentists have alternatives to amalgam. New composites, resins, and porcelains are being used to repair decayed, broken, or missing teeth. These new materials are strong, chemically resistant to body fluids, and can be altered to have the natural color of the tooth. Some of the new resins also contain fluoride that will protect the tooth from further decay. Many of these new materials would be useless without the development of new bonding agents. The new "glues" or bonding agents adhere the new materials to the natural tooth. These bonding agents must be strong and chemically resistant to body fluids.

Reading Check *Why are these new dental materials desirable for repairing teeth?*

Orthodontists are using new nickel and titanium alloys for the wires on braces. These wires have shape memory. The wires undergo a special heat treatment process to lock in their shapes. If the wires are forced out of their heat-treated shape, the wires will try to return to their original shape. Orthodontists have found these wires useful in straightening crooked teeth. How do you think these wires help to straighten crooked teeth?

section 3 review

Summary

Transition Elements

- Groups 3–12, which are transition elements, are all metals.
- Their properties change less than the representative elements.
- The elements in the iron triad are iron, cobalt, and nickel.

Inner Transition Elements

- The lanthanide series contains elements from cerium to lutetium.
- The lanthanides also are known as the rare earth elements.
- The actinide series contains elements from thorium to lawrencium.

Self Check

1. **State** how the elements in the iron triad differ from other transition metals.
2. **Explain** the major difference between the lanthanides and actinides.
3. **Explain** how mercury is used.
4. **Describe** how synthetic elements are made.
5. **Think Critically** Iridium and cadmium are both transition elements. Predict which element is toxic and which element is more likely to be a catalyst. Explain.

Applying Skills

6. **Form Hypotheses** How does the appearance of a burned-out lightbulb compare to a new lightbulb? What could explain the difference?

Metals and Nonmetals

Real-World Question

Metals on asteroids appear attractive for mining to space programs because the metals are essential for space travel. An asteroid could be processed to provide very pure iron and nickel. Valuable by-products would include cobalt, platinum, and gold. How can miners determine if an element is a metal or a nonmetal?

Goals

■ **Describe** the appearance of metals and nonmetals.

■ **Evaluate** the malleability or brittleness of metals and nonmetals

■ **Observe** chemical reactions of metals and nonmetals with an acid and a base.

Materials (per group of 2–3 students)

10 test tubes with rack
10-mL graduated cylinder marking pencil
forceps or tweezers 25 g carbon
small hammer or mallet 25 g silicon
dropper bottle of 0.5*M* HCl 25 g tin
dropper bottle of 0.1*M* CuCl₂ 25 g sulfur
test-tube brush 25 g iron

Safety Precautions

Procedure

1. Copy data table into your Science Journal. Fill in data table as you complete the lab.

2. Describe in as much detail as possible the appearance of the sample, including color, luster, and state of matter.

3. Use the hammer or mallet to determine malleability or brittleness.

4. Label 5 test tubes #1–5. Place a 1-g sample of each element in separate test tubes. Add 5 mL of HCl to each tube. If bubbles form, this indicates a chemical reaction.

5. Repeat step 4, substituting HCl with CuCl₂. Do not discard the solutions immediately. Continue to observe for five minutes. Some of the changes may be slow. A chemical reaction is indicated by a change in appearance of the element.

Analyze Your Data

1. **Analyze Results** What characteristics distinguish metals from nonmetals?

2. **List** which elements you discovered to be metals.

3. **Describe** a metalloid. Are any of the elements tested a metalloid? If so, name them.

Conclude and Apply

1. **Explain** how the future might increase or decrease the need for selected elements.

2. **Infer** why discovering and mining metals on asteroids might be an important find.

Metal and Nonmetal Data				
Element	Appearance	Malleable or Brittle	Reaction wth HCl	Reaction with CuCl₂
Carbon				
Silicon				
Tin	Do not write in this book.			
Sulfur				
Iron				

Use the Internet

Health Risks from Heavy Metals

Goals

■ **Organize** and synthesize information on a chemical or heavy metal thought to cause health problems in the area where you live.

■ **Communicate** your findings to others in your class.

Data Source

Science Online

Visit bookk.msscience.com/ internet_lab for more information about health risks from heavy metals, hints on health risks, and data from other students.

◉ Real-World Question

Many heavy metals are found naturally on the planet. People and animals are exposed to these metals every day. One way to reduce the exposure is to know as much as possible about the effects of chemicals on you and the environment. Do heavy metals and other chemicals pose a threat to the health of humans? Could health problems be caused by exposure to heavy metals such as lead, or a radioactive chemical element, such as radon? Is the incidence of these problems higher in one area than another?

◉ Make a Plan

1. Read general information concerning heavy metals and other potentially hazardous chemicals.

2. Use the sites listed at the link to the left to research possible health problems in your area caused by exposure to chemicals or heavy metals. Do you see a pattern in the type of health risks that you found in your research?

3. Check the link to the left to see what others have learned.

Health Risk Data Table				
Location	Chemical or Heavy Metal	How People Come in Contact with Chemical	Potential Health Problem	Who Is Affected
		Do not write in this book.		

▶ Follow Your Plan

1. Make sure your teacher approves your plan before you start.

2. **Research information** that can help you find out about health risks in your area.

3. **Organize** your information in a data table like the one shown.

4. **Write** a report in your Science Journal using the results of your research on heavy metals.

5. Post your data in the table provided at the link below.

▶ Analyze Your Data

1. **Evaluate** Did all your sources agree on the health risk of the chemical or heavy metal?

2. **Analyze** all your sources for possible bias. Are some sources more reliable than others?

3. **Explain** how the health risk differs for adults and children.

4. **Identify** the sources of the heavy metals in your area. Are the heavy metals still being deposited in your area?

▶ Conclude and Apply

1. **Analyze Results** Were the same substances found to be health risks in other parts of the country? From the data at the link below, try to predict what chemicals or heavy metals are health risks in different parts of the country.

2. **Determine** what information you think is the most important for the public to be aware of.

3. **Explain** what could be done to decrease the risk of the health problems you identified.

*C*ommunicating
Your Data

Find this lab using the link below. **Post** your data in the table provided. **Compare** your data to those of other students. **Analyze** and look for patterns in the data.

Science Online

bookk.msscience.com/internet_lab

Anansi Tries to Steal All the Wisdom in the World
A folktale, adapted by Matt Evans

The following African folktale about a spider named Anansi (or Anancy) is from the Ashanti people in Western Africa.

Anansi the spider knew that he was not wise… "I know… if I can get all of the wisdom in the village and put it in a hollow gourd… I would be the wisest of all!" So he set out to find a suitable gourd and then began his journey to collect the village's wisdom… He looked around and spotted a tall, tall tree. "Ah," he said to himself, "if I could hide my wisdom high in that tree, I would never have to worry about someone stealing it from me!"… He first took a cloth band and tied it around his waist. Then he tied the heavy gourd to the front of his belly where it would be safe. As he began to climb, however, the gourd full of wisdom kept getting in the way…

Soon Anansi's youngest son walked by… "But Father," said the son, "wouldn't it be much easier if you tied the gourd behind you instead of in front?"… Anansi moved the gourd so that it was behind him and proceeded up the tree with no problems at all. When he had reached the top, he cried out, "I walked all over and collected so much wisdom I am the wisest person ever, but still my baby son is wiser than me. Take back your wisdom!" He lifted the gourd high over his head and spilled its contents into the wind. The wisdom blew far and wide and settled across the land. And this is how wisdom came to the world.

Understanding Literature

Folktales The African folktale you have just read is called an animal-trickster tale. Trickster tales come from Africa, the Caribbean, and Latin American countries. Trickster tales portray a wily and cunning animal or human who at times bewilders the more powerful and at other times becomes a victim of his or her own schemes. Describe other kinds of folktales, such as fairy tales and tall tales

Respond to the Reading

1. Is Anansi a clever spider?
2. Why did Anansi scatter the wisdom he had collected?
3. **Linking Science and Writing** Write a folktale featuring an animal as a trickster.

Elements are classified in relation to one another in a periodic table. They also are classified in groups of elements that share similar characteristics. Thus, there is a group of elements known as the alkali metals, another called the halogens, and so on. This way of classifying elements is similar to the way in which folktales are classified. Trickster tales have similar characteristics such as a character that has certain traits, like cleverness, wit, cunning, and an ability to survive.

Reviewing Main Ideas

Section 1 Introduction to the Periodic Table

1. When organized according to atomic number in a table, elements with similar properties occupy the same column and are called a group or family.

2. On the periodic table, the properties of the elements change gradually across a horizontal row called a period.

3. The periodic table can be divided into representative elements and transition elements.

Section 2 Representative Elements

1. The groups on the periodic table are known also by other names. For instance, Group 17 is known as halogens.

2. Atoms of elements in Groups 1 and 2 readily combine with atoms of other elements.

3. Each element in Group 2 combines less readily than its neighbor in Group 1. Each alkaline earth metal is denser and has a higher melting point than the alkali metal in its period.

4. Sodium, potassium, magnesium, and calcium have important biological roles.

Section 3 Transition Elements

1. The metals in the iron triad are found in a variety of places. Iron is found in blood and in the structure of skyscrapers.

2. Copper, silver, and gold are fairly unreactive, malleable elements.

3. The lanthanides are naturally occurring elements with similar properties.

4. The actinides are radioactive elements. All actinides except thorium, proactinium, and uranium are synthetic.

Visualizing Main Ideas

Copy and complete the following concept map on the periodic table.

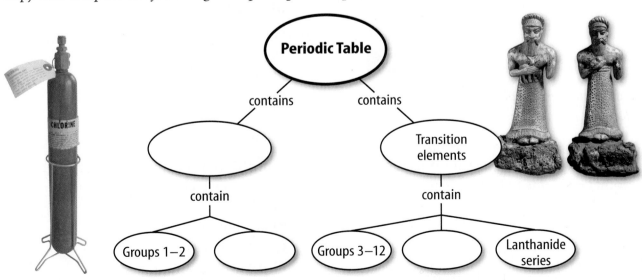

Using Vocabulary

actinides p. 114
alkali metals p. 105
alkaline earth
   metals p. 106
catalyst p. 113
group p. 99
halogens p. 110
lanthanides p. 114
metal p. 102

metalloid p. 102
noble gases p. 110
nonmetal p. 102
period p. 99
representative
   element p. 99
semiconductor p. 107
synthetic elements p. 114
transition elements p. 99

Answer the following questions using complete sentences.

1. What is the difference between a group and a period?

2. What is the connection between a metalloid and a semiconductor?

3. What is a catalyst?

4. Arrange the terms *nonmetal, metal,* and *metalloid* according to increasing heat and electrical conductivity.

5. How is a metalloid like a metal? How is it different from a metal?

6. What are synthetic elements?

7. How are transition elements alike?

8. Why are some gases considered to be noble?

Checking Concepts

Choose the word or phrase that best answers the question.

9. Which of the following groups from the periodic table combines most readily with other elements to form compounds?
   A) transition metals
   B) alkaline earth metals
   C) alkali metals
   D) iron triad

10. Which element is NOT a part of the iron triad?
    A) nickel
    B) copper
    C) cobalt
    D) iron

11. Which element is located in Group 6, period 4?
    A) tungsten   C) titanium
    B) chromium   D) hafnium

12. Which element below is NOT a transition element?
    A) gold   C) silver
    B) calcium   D) copper

13. Several groups contain only metals. Which group contains only nonmetals?
    A) Group 1   C) Group 2
    B) Group 12   D) Group 18

14. Which of the following elements is likely to be contained in a substance with a brilliant yellow color?
    A) chromium   C) iron
    B) carbon   D) tin

15. Which halogen is radioactive?
    A) astatine   C) bromine
    B) chlorine   D) iodine

16. Which of the following describes the element tellurium?
    A) alkali metal
    B) transition metal
    C) metalloid
    D) lanthanide

17. A brittle, non-conducting, solid might belong to which of the following groups?
    A) alkali metals
    B) alkaline earth metals
    C) actinide series
    D) oxygen group

 Science Online bookk.msscience.com/vocabulary_puzzlemaker

Thinking Critically

18. Explain why it is important that mercury be kept out of streams and waterways.

19. Determine If you were going to try to get the noble gas argon to combine with another element, would fluorine be a good choice for the other element? Explain.

Use the figure below to answer question 20.

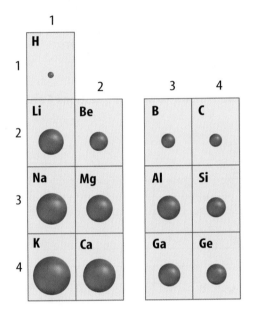

20. Interpret Data The periodic table shows trends across the rows and down the columns. In this portion of the periodic table, the relative size of the atom is represented by a ball. What trend can you see in this part of the table for relative size?

21. Evaluate It is theorized that some of the actinides beyond uranium were once present in Earth's crust. If this theory is true, how would their half-lives compare with the half-life of uranium, which is 4.5 billion years?

22. Recognize Cause and Effect Why do photographers work in low light when they work with materials containing selenium?

23. Predict How would life on Earth be different if the atmosphere were 80 percent oxygen and 20 percent nitrogen instead of the other way around?

24. Compare and contrast Na and Mg, which are in the same period, with F and Cl, which are in the same group.

Performance Activities

25. Ask Questions Research the contribution that Henry G. J. Moseley made to the development of the modern periodic table. Research the background and work of this scientist. Write your findings in the form of an interview.

Applying Math

26. Elements at Room Temperature Make a bar graph of the representative elements that shows how many of the elements are solids, liquids, and gases at room temperature.

27. Calculate Using the information that you collected in question 26, calculate the percentage of solids, liquids, and gases within the representative elements.

Use the figure below to answer question 28.

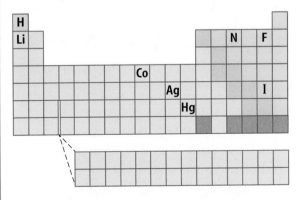

28. Element Details For each element shown, give the element's period and group number; whether the element is a metal or a nonmetal; and whether it is a solid, liquid, or gas at room temperature.

Part 1 | Multiple Choice

Record your answers on the answer sheet provided by your teacher or on a sheet of paper.

1. Which statement about the periodic table is TRUE?
   A. Elements all occur naturally on Earth.
   B. Elements occur in the order in which they were discovered.
   C. Elements with similar properties occupy the same group.
   D. Elements are arranged in the order Mendeleev chose.

2. Which of these is NOT a property of metals?
   A. malleability
   B. luster
   C. ductility
   D. poor conductor of heat and electricity

Use the illustration below to answer questions 3 and 4.

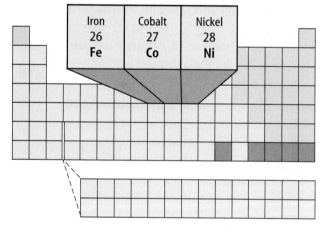

Iron	Cobalt	Nickel
26	27	28
Fe	Co	Ni

3. What name is given to these three elements which are used in processes that create steel and other metal mixtures?
   A. lanthanides   C. actinides
   B. the coin metals   D. the iron triad

Test-Taking Tip

The Best Answer Read all choices before answering the questions.

4. To which category do these elements belong?
   A. nonmetals
   B. transition elements
   C. noble gases
   D. representative metals

5. Which member of the boron family is used to make soft-drink cans, baseball bats, and siding for homes?
   A. aluminum   C. indium
   B. boron   D. gallium

Use the table below to answer questions 6 and 7.

H																	He
Li	Be														F	Ne
Na	Mg														Cl	Ar
K	Ca														Br	Kr
Rb	Sr														I	Xe
Cs	Ba														At	Rn
Fr	Ra															

6. Halogens are highly reactive nonmetals. Which group combines most readily with them?
   A. Group 1, alkali metals
   B. Group 2, alkaline earth metals
   C. Group 17, halogens
   D. Group 18, noble gases

7. Which alkali metal element is most reactive?
   A. Li   C. K
   B. Na   D. Cs

8. Many elements that are essential for life, including nitrogen, oxygen, and carbon, are part of what classification?
   A. nonmetals   C. metalloids
   B. metals   D. noble gases

Part 2 | Short Response/Grid In

Record your answers on the answer sheet provided by your teacher or on a sheet of paper.

9. Based on the information found in the periodic table, compare and contrast properties of the elements gold and silver.

10. Why don't the element symbols always match the name? Give two examples and describe the origin of each symbol.

Use the graph below to answer questions 11 and 12.

Boiling Points of Period 1, 2, and 3 Elements

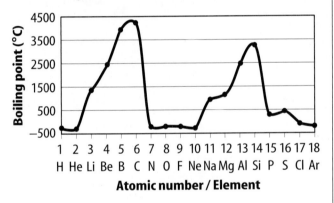

11. The data shows that boiling point is a periodic property. Explain what the term *periodic property* means.

12. Describe patterns evident in this data.

13. Describe the mixture used by dentists for the past 150 years to fill cavities in decayed teeth. Why do many dentists today use other materials to repair teeth?

14. Compare and contrast the periodic table that Mendeleev developed to the periodic table that Mosley organized.

15. Choose a representative element group and list the elements in that group. Then list three to four uses for those elements.

Part 3 | Open Ended

Record your answers on a sheet of paper.

16. What role does nitrogen play in the human body? Explain the importance of bacteria in the soil which change the form in which nitrogen naturally occurs.

17. Much of the wiring in houses is made from copper. What properties of copper make it ideal for this purpose?

18. Why do some homeowners check for the presence of the noble gas radon in their homes?

Use the graph below to answer questions 19 and 20.

Elements in the Human Body

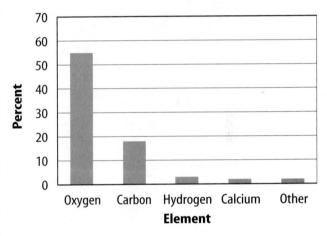

19. The graph above shows elements present in the greatest amounts in the human body. Use information from the periodic table to create a chart which shows properties of each element, including its symbol, atomic number, the group to which it belongs, and whether it is a metal, non-metal, or metalloid.

20. One element shown here is an alkaline earth metal. Compare the properties of the elements in this family to those of the elements found in Group 1.

Student Resources

CONTENTS

Science Skill Handbook 128

Scientific Methods 128
Identify a Question 128
Gather and Organize
  Information 128
Form a Hypothesis 131
Test the Hypothesis 132
Collect Data 132
Analyze the Data 135
Draw Conclusions 136
Communicate 136
Safety Symbols 137
Safety in the Science Laboratory 138
General Safety Rules 138
Prevent Accidents 138
Laboratory Work 138
Laboratory 139
Emergencies 139

Extra Try at Home Labs 140

Comparing Particles 140
Microscopic Crystals 140
Good and Bad Apples 141
Research Race 141

Technology Skill Handbook ...142

Computer Skills 142
Use a Word Processing Program ...142
Use a Database 143
Use the Internet 143
Use a Spreadsheet 144
Use Graphics Software 144
Presentation Skills 145
Develop Multimedia
  Presentations 145
Computer Presentations 145

Math Skill Handbook 146

Math Review 146
Use Fractions 146
Use Ratios 149
Use Decimals 150
Use Proportions 150
Use Percentages 151
Solve One-Step Equations 151
Use Statistics 152
Use Geometry 153
Science Applications 156
Measure in SI 156
Dimensional Analysis 156
Precision and Significant Digits ...158
Scientific Notation 158
Make and Use Graphs 159

Reference Handbooks 161

Physical Science Reference Tables 161
Periodic Table of the Elements 162
Physical Science References 164

English/Spanish Glossary 165

Index 170

Credits 174

Scientific Methods

Scientists use an orderly approach called the scientific method to solve problems. This includes organizing and recording data so others can understand them. Scientists use many variations in this method when they solve problems.

Identify a Question

The first step in a scientific investigation or experiment is to identify a question to be answered or a problem to be solved. For example, you might ask which gasoline is the most efficient.

Gather and Organize Information

After you have identified your question, begin gathering and organizing information. There are many ways to gather information, such as researching in a library, interviewing those knowledgeable about the subject, testing and working in the laboratory and field. Fieldwork is investigations and observations done outside of a laboratory.

Researching Information Before moving in a new direction, it is important to gather the information that already is known about the subject. Start by asking yourself questions to determine exactly what you need to know. Then you will look for the information in various reference sources, like the student is doing in **Figure 1.** Some sources may include textbooks, encyclopedias, government documents, professional journals, science magazines, and the Internet. Always list the sources of your information.

Figure 1 The Internet can be a valuable research tool.

Evaluate Sources of Information Not all sources of information are reliable. You should evaluate all of your sources of information, and use only those you know to be dependable. For example, if you are researching ways to make homes more energy efficient, a site written by the U.S. Department of Energy would be more reliable than a site written by a company that is trying to sell a new type of weatherproofing material. Also, remember that research always is changing. Consult the most current resources available to you. For example, a 1985 resource about saving energy would not reflect the most recent findings.

Sometimes scientists use data that they did not collect themselves, or conclusions drawn by other researchers. This data must be evaluated carefully. Ask questions about how the data were obtained, if the investigation was carried out properly, and if it has been duplicated exactly with the same results. Would you reach the same conclusion from the data? Only when you have confidence in the data can you believe it is true and feel comfortable using it.

Interpret Scientific Illustrations As you research a topic in science, you will see drawings, diagrams, and photographs to help you understand what you read. Some illustrations are included to help you understand an idea that you can't see easily by yourself, like the tiny particles in an atom in **Figure 2.** A drawing helps many people to remember details more easily and provides examples that clarify difficult concepts or give additional information about the topic you are studying. Most illustrations have labels or a caption to identify or to provide more information.

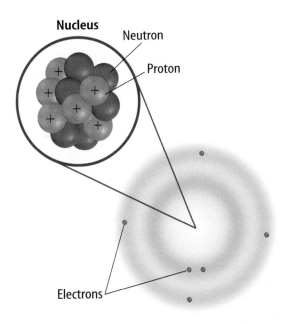

Figure 2 This drawing shows an atom of carbon with its six protons, six neutrons, and six electrons.

Concept Maps One way to organize data is to draw a diagram that shows relationships among ideas (or concepts). A concept map can help make the meanings of ideas and terms more clear, and help you understand and remember what you are studying. Concept maps are useful for breaking large concepts down into smaller parts, making learning easier.

Network Tree A type of concept map that not only shows a relationship, but how the concepts are related is a network tree, shown in **Figure 3.** In a network tree, the words are written in the ovals, while the description of the type of relationship is written across the connecting lines.

When constructing a network tree, write down the topic and all major topics on separate pieces of paper or notecards. Then arrange them in order from general to specific. Branch the related concepts from the major concept and describe the relationship on the connecting line. Continue to more specific concepts until finished.

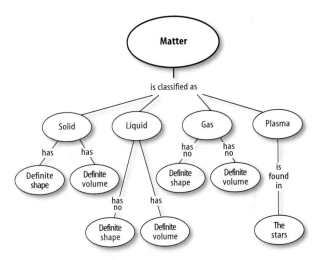

Figure 3 A network tree shows how concepts or objects are related.

Events Chain Another type of concept map is an events chain. Sometimes called a flow chart, it models the order or sequence of items. An events chain can be used to describe a sequence of events, the steps in a procedure, or the stages of a process.

When making an events chain, first find the one event that starts the chain. This event is called the initiating event. Then, find the next event and continue until the outcome is reached, as shown in **Figure 4.**

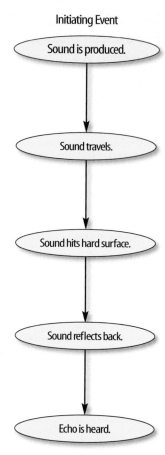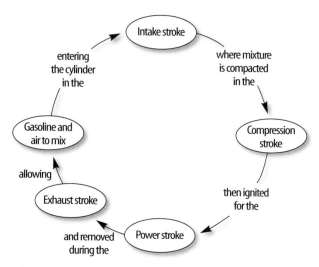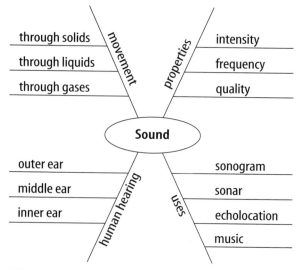

Initiating Event

Figure 4 Events-chain concept maps show the order of steps in a process or event. This concept map shows how a sound makes an echo.

Figure 5 A cycle map shows events that occur in a cycle.

Cycle Map A specific type of events chain is a cycle map. It is used when the series of events do not produce a final outcome, but instead relate back to the beginning event, such as in **Figure 5.** Therefore, the cycle repeats itself.

To make a cycle map, first decide what event is the beginning event. This is also called the initiating event. Then list the next events in the order that they occur, with the last event relating back to the initiating event. Words can be written between the events that describe what happens from one event to the next. The number of events in a cycle map can vary, but usually contain three or more events.

Spider Map A type of concept map that you can use for brainstorming is the spider map. When you have a central idea, you might find that you have a jumble of ideas that relate to it but are not necessarily clearly related to each other. The spider map on sound in **Figure 6** shows that if you write these ideas outside the main concept, then you can begin to separate and group unrelated terms so they become more useful.

Figure 6 A spider map allows you to list ideas that relate to a central topic but not necessarily to one another.

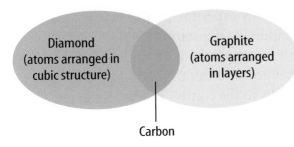

Figure 7 This Venn diagram compares and contrasts two substances made from carbon.

Venn Diagram To illustrate how two subjects compare and contrast you can use a Venn diagram. You can see the characteristics that the subjects have in common and those that they do not, shown in **Figure 7.**

To create a Venn diagram, draw two overlapping ovals that that are big enough to write in. List the characteristics unique to one subject in one oval, and the characteristics of the other subject in the other oval. The characteristics in common are listed in the overlapping section.

Make and Use Tables One way to organize information so it is easier to understand is to use a table. Tables can contain numbers, words, or both.

To make a table, list the items to be compared in the first column and the characteristics to be compared in the first row. The title should clearly indicate the content of the table, and the column or row heads should be clear. Notice that in **Table 1** the units are included.

Table 1 Recyclables Collected During Week			
Day of Week	**Paper (kg)**	**Aluminum (kg)**	**Glass (kg)**
Monday	5.0	4.0	12.0
Wednesday	4.0	1.0	10.0
Friday	2.5	2.0	10.0

Make a Model One way to help you better understand the parts of a structure, the way a process works, or to show things too large or small for viewing is to make a model. For example, an atomic model made of a plastic-ball nucleus and pipe-cleaner electron shells can help you visualize how the parts of an atom relate to each other. Other types of models can by devised on a computer or represented by equations.

Form a Hypothesis

A possible explanation based on previous knowledge and observations is called a hypothesis. After researching gasoline types and recalling previous experiences in your family's car you form a hypothesis—our car runs more efficiently because we use premium gasoline. To be valid, a hypothesis has to be something you can test by using an investigation.

Predict When you apply a hypothesis to a specific situation, you predict something about that situation. A prediction makes a statement in advance, based on prior observation, experience, or scientific reasoning. People use predictions to make everyday decisions. Scientists test predictions by performing investigations. Based on previous observations and experiences, you might form a prediction that cars are more efficient with premium gasoline. The prediction can be tested in an investigation.

Design an Experiment A scientist needs to make many decisions before beginning an investigation. Some of these include: how to carry out the investigation, what steps to follow, how to record the data, and how the investigation will answer the question. It also is important to address any safety concerns.

Test the Hypothesis

Now that you have formed your hypothesis, you need to test it. Using an investigation, you will make observations and collect data, or information. This data might either support or not support your hypothesis. Scientists collect and organize data as numbers and descriptions.

Follow a Procedure In order to know what materials to use, as well as how and in what order to use them, you must follow a procedure. **Figure 8** shows a procedure you might follow to test your hypothesis.

Procedure
1. Use regular gasoline for two weeks.
2. Record the number of kilometers between fill-ups and the amount of gasoline used.
3. Switch to premium gasoline for two weeks.
4. Record the number of kilometers between fill-ups and the amount of gasoline used.

Figure 8 A procedure tells you what to do step by step.

Identify and Manipulate Variables and Controls In any experiment, it is important to keep everything the same except for the item you are testing. The one factor you change is called the independent variable. The change that results is the dependent variable. Make sure you have only one independent variable, to assure yourself of the cause of the changes you observe in the dependent variable. For example, in your gasoline experiment the type of fuel is the independent variable. The dependent variable is the efficiency.

Many experiments also have a control—an individual instance or experimental subject for which the independent variable is not changed. You can then compare the test results to the control results. To design a control you can have two cars of the same type. The control car uses regular gasoline for four weeks. After you are done with the test, you can compare the experimental results to the control results.

Collect Data

Whether you are carrying out an investigation or a short observational experiment, you will collect data, as shown in **Figure 9.** Scientists collect data as numbers and descriptions and organize it in specific ways.

Observe Scientists observe items and events, then record what they see. When they use only words to describe an observation, it is called qualitative data. Scientists' observations also can describe how much there is of something. These observations use numbers, as well as words, in the description and are called quantitative data. For example, if a sample of the element gold is described as being "shiny and very dense" the data are qualitative. Quantitative data on this sample of gold might include "a mass of 30 g and a density of 19.3 g/cm^3."

Figure 9 Collecting data is one way to gather information directly.

Figure 10 Record data neatly and clearly so it is easy to understand.

When you make observations you should examine the entire object or situation first, and then look carefully for details. It is important to record observations accurately and completely. Always record your notes immediately as you make them, so you do not miss details or make a mistake when recording results from memory. Never put unidentified observations on scraps of paper. Instead they should be recorded in a notebook, like the one in **Figure 10.** Write your data neatly so you can easily read it later. At each point in the experiment, record your observations and label them. That way, you will not have to determine what the figures mean when you look at your notes later. Set up any tables that you will need to use ahead of time, so you can record any observations right away. Remember to avoid bias when collecting data by not including personal thoughts when you record observations. Record only what you observe.

Estimate Scientific work also involves estimating. To estimate is to make a judgment about the size or the number of something without measuring or counting. This is important when the number or size of an object or population is too large or too difficult to accurately count or measure.

Sample Scientists may use a sample or a portion of the total number as a type of estimation. To sample is to take a small, representative portion of the objects or organisms of a population for research. By making careful observations or manipulating variables within that portion of the group, information is discovered and conclusions are drawn that might apply to the whole population. A poorly chosen sample can be unrepresentative of the whole. If you were trying to determine the rainfall in an area, it would not be best to take a rainfall sample from under a tree.

Measure You use measurements everyday. Scientists also take measurements when collecting data. When taking measurements, it is important to know how to use measuring tools properly. Accuracy also is important.

Length To measure length, the distance between two points, scientists use meters. Smaller measurements might be measured in centimeters or millimeters.

Length is measured using a metric ruler or meter stick. When using a metric ruler, line up the 0-cm mark with the end of the object being measured and read the number of the unit where the object ends. Look at the metric ruler shown in **Figure 11.** The centimeter lines are the long, numbered lines, and the shorter lines are millimeter lines. In this instance, the length would be 4.50 cm.

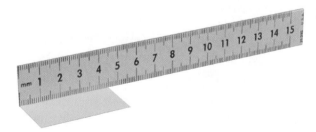

Figure 11 This metric ruler has centimeter and millimeter divisions.

Mass The SI unit for mass is the kilogram (kg). Scientists can measure mass using units formed by adding metric prefixes to the unit gram (g), such as milligram (mg). To measure mass, you might use a triple-beam balance similar to the one shown in **Figure 12.** The balance has a pan on one side and a set of beams on the other side. Each beam has a rider that slides on the beam.

When using a triple-beam balance, place an object on the pan. Slide the largest rider along its beam until the pointer drops below zero. Then move it back one notch. Repeat the process for each rider proceeding from the larger to smaller until the pointer swings an equal distance above and below the zero point. Sum the masses on each beam to find the mass of the object. Move all riders back to zero when finished.

Instead of putting materials directly on the balance, scientists often take a tare of a container. A tare is the mass of a container into which objects or substances are placed for measuring their masses. To mass objects or substances, find the mass of a clean container. Remove the container from the pan, and place the object or substances in the container. Find the mass of the container with the materials in it. Subtract the mass of the empty container from the mass of the filled container to find the mass of the materials you are using.

Figure 12 A triple-beam balance is used to determine the mass of an object.

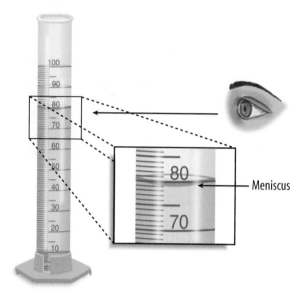

Meniscus

Figure 13 Graduated cylinders measure liquid volume.

Liquid Volume To measure liquids, the unit used is the liter. When a smaller unit is needed, scientists might use a milliliter. Because a milliliter takes up the volume of a cube measuring 1 cm on each side it also can be called a cubic centimeter ($cm^3 = cm \times cm \times cm$).

You can use beakers and graduated cylinders to measure liquid volume. A graduated cylinder, shown in **Figure 13,** is marked from bottom to top in milliliters. In lab, you might use a 10-mL graduated cylinder or a 100-mL graduated cylinder. When measuring liquids, notice that the liquid has a curved surface. Look at the surface at eye level, and measure the bottom of the curve. This is called the meniscus. The graduated cylinder in **Figure 13** contains 79.0 mL, or 79.0 cm^3, of a liquid.

Temperature Scientists often measure temperature using the Celsius scale. Pure water has a freezing point of 0°C and boiling point of 100°C. The unit of measurement is degrees Celsius. Two other scales often used are the Fahrenheit and Kelvin scales.

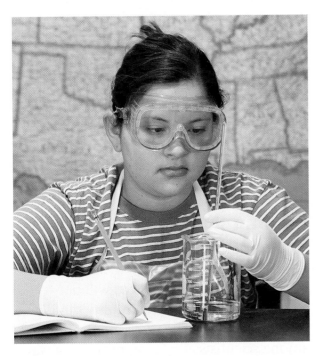

Figure 14 A thermometer measures the temperature of an object.

Scientists use a thermometer to measure temperature. Most thermometers in a laboratory are glass tubes with a bulb at the bottom end containing a liquid such as colored alcohol. The liquid rises or falls with a change in temperature. To read a glass thermometer like the thermometer in **Figure 14,** rotate it slowly until a red line appears. Read the temperature where the red line ends.

Form Operational Definitions An operational definition defines an object by how it functions, works, or behaves. For example, when you are playing hide and seek and a tree is home base, you have created an operational definition for a tree.

Objects can have more than one operational definition. For example, a ruler can be defined as a tool that measures the length of an object (how it is used). It can also be a tool with a series of marks used as a standard when measuring (how it works).

Analyze the Data

To determine the meaning of your observations and investigation results, you will need to look for patterns in the data. Then you must think critically to determine what the data mean. Scientists use several approaches when they analyze the data they have collected and recorded. Each approach is useful for identifying specific patterns.

Interpret Data The word *interpret* means "to explain the meaning of something." When analyzing data from an experiment, try to find out what the data show. Identify the control group and the test group to see whether or not changes in the independent variable have had an effect. Look for differences in the dependent variable between the control and test groups.

Classify Sorting objects or events into groups based on common features is called classifying. When classifying, first observe the objects or events to be classified. Then select one feature that is shared by some members in the group, but not by all. Place those members that share that feature in a subgroup. You can classify members into smaller and smaller subgroups based on characteristics. Remember that when you classify, you are grouping objects or events for a purpose. Keep your purpose in mind as you select the features to form groups and subgroups.

Compare and Contrast Observations can be analyzed by noting the similarities and differences between two more objects or events that you observe. When you look at objects or events to see how they are similar, you are comparing them. Contrasting is looking for differences in objects or events.

Recognize Cause and Effect A cause is a reason for an action or condition. The effect is that action or condition. When two events happen together, it is not necessarily true that one event caused the other. Scientists must design a controlled investigation to recognize the exact cause and effect.

Draw Conclusions

When scientists have analyzed the data they collected, they proceed to draw conclusions about the data. These conclusions are sometimes stated in words similar to the hypothesis that you formed earlier. They may confirm a hypothesis, or lead you to a new hypothesis.

Infer Scientists often make inferences based on their observations. An inference is an attempt to explain observations or to indicate a cause. An inference is not a fact, but a logical conclusion that needs further investigation. For example, you may infer that a fire has caused smoke. Until you investigate, however, you do not know for sure.

Apply When you draw a conclusion, you must apply those conclusions to determine whether the data supports the hypothesis. If your data do not support your hypothesis, it does not mean that the hypothesis is wrong. It means only that the result of the investigation did not support the hypothesis. Maybe the experiment needs to be redesigned, or some of the initial observations on which the hypothesis was based were incomplete or biased. Perhaps more observation or research is needed to refine your hypothesis. A successful investigation does not always come out the way you originally predicted.

Avoid Bias Sometimes a scientific investigation involves making judgments. When you make a judgment, you form an opinion. It is important to be honest and not to allow any expectations of results to bias your judgments. This is important throughout the entire investigation, from researching to collecting data to drawing conclusions.

Communicate

The communication of ideas is an important part of the work of scientists. A discovery that is not reported will not advance the scientific community's understanding or knowledge. Communication among scientists also is important as a way of improving their investigations.

Scientists communicate in many ways, from writing articles in journals and magazines that explain their investigations and experiments, to announcing important discoveries on television and radio. Scientists also share ideas with colleagues on the Internet or present them as lectures, like the student is doing in **Figure 15.**

Figure 15 A student communicates to his peers about his investigation.

SAFETY SYMBOLS	HAZARD	EXAMPLES	PRECAUTION	REMEDY
DISPOSAL	Special disposal procedures need to be followed.	certain chemicals, living organisms	Do not dispose of these materials in the sink or trash can.	Dispose of wastes as directed by your teacher.
BIOLOGICAL	Organisms or other biological materials that might be harmful to humans	bacteria, fungi, blood, unpreserved tissues, plant materials	Avoid skin contact with these materials. Wear mask or gloves.	Notify your teacher if you suspect contact with material. Wash hands thoroughly.
EXTREME TEMPERATURE	Objects that can burn skin by being too cold or too hot	boiling liquids, hot plates, dry ice, liquid nitrogen	Use proper protection when handling.	Go to your teacher for first aid.
SHARP OBJECT	Use of tools or glassware that can easily puncture or slice skin	razor blades, pins, scalpels, pointed tools, dissecting probes, broken glass	Practice common-sense behavior and follow guidelines for use of the tool.	Go to your teacher for first aid.
FUME	Possible danger to respiratory tract from fumes	ammonia, acetone, nail polish remover, heated sulfur, moth balls	Make sure there is good ventilation. Never smell fumes directly. Wear a mask.	Leave foul area and notify your teacher immediately.
ELECTRICAL	Possible danger from electrical shock or burn	improper grounding, liquid spills, short circuits, exposed wires	Double-check setup with teacher. Check condition of wires and apparatus.	Do not attempt to fix electrical problems. Notify your teacher immediately.
IRRITANT	Substances that can irritate the skin or mucous membranes of the respiratory tract	pollen, moth balls, steel wool, fiberglass, potassium permanganate	Wear dust mask and gloves. Practice extra care when handling these materials.	Go to your teacher for first aid.
CHEMICAL	Chemicals can react with and destroy tissue and other materials	bleaches such as hydrogen peroxide; acids such as sulfuric acid, hydrochloric acid; bases such as ammonia, sodium hydroxide	Wear goggles, gloves, and an apron.	Immediately flush the affected area with water and notify your teacher.
TOXIC	Substance may be poisonous if touched, inhaled, or swallowed.	mercury, many metal compounds, iodine, poinsettia plant parts	Follow your teacher's instructions.	Always wash hands thoroughly after use. Go to your teacher for first aid.
FLAMMABLE	Flammable chemicals may be ignited by open flame, spark, or exposed heat.	alcohol, kerosene, potassium permanganate	Avoid open flames and heat when using flammable chemicals.	Notify your teacher immediately. Use fire safety equipment if applicable.
OPEN FLAME	Open flame in use, may cause fire.	hair, clothing, paper, synthetic materials	Tie back hair and loose clothing. Follow teacher's instruction on lighting and extinguishing flames.	Notify your teacher immediately. Use fire safety equipment if applicable.

 Eye Safety Proper eye protection should be worn at all times by anyone performing or observing science activities.

 Clothing Protection This symbol appears when substances could stain or burn clothing.

 Animal Safety This symbol appears when safety of animals and students must be ensured.

 Handwashing After the lab, wash hands with soap and water before removing goggles.

Safety in the Science Laboratory

The science laboratory is a safe place to work if you follow standard safety procedures. Being responsible for your own safety helps to make the entire laboratory a safer place for everyone. When performing any lab, read and apply the caution statements and safety symbol listed at the beginning of the lab.

General Safety Rules

1. Obtain your teacher's permission to begin all investigations and use laboratory equipment.

2. Study the procedure. Ask your teacher any questions. Be sure you understand safety symbols shown on the page.

3. Notify your teacher about allergies or other health conditions which can affect your participation in a lab.

4. Learn and follow use and safety procedures for your equipment. If unsure, ask your teacher.

5. Never eat, drink, chew gum, apply cosmetics, or do any personal grooming in the lab. Never use lab glassware as food or drink containers. Keep your hands away from your face and mouth.

6. Know the location and proper use of the safety shower, eye wash, fire blanket, and fire alarm.

Prevent Accidents

1. Use the safety equipment provided to you. Goggles and a safety apron should be worn during investigations.

2. Do NOT use hair spray, mousse, or other flammable hair products. Tie back long hair and tie down loose clothing.

3. Do NOT wear sandals or other open-toed shoes in the lab.

4. Remove jewelry on hands and wrists. Loose jewelry, such as chains and long necklaces, should be removed to prevent them from getting caught in equipment.

5. Do not taste any substances or draw any material into a tube with your mouth.

6. Proper behavior is expected in the lab. Practical jokes and fooling around can lead to accidents and injury.

7. Keep your work area uncluttered.

Laboratory Work

1. Collect and carry all equipment and materials to your work area before beginning a lab.

2. Remain in your own work area unless given permission by your teacher to leave it.

3. Always slant test tubes away from yourself and others when heating them, adding substances to them, or rinsing them.

4. If instructed to smell a substance in a container, hold the container a short distance away and fan vapors towards your nose.

5. Do NOT substitute other chemicals/substances for those in the materials list unless instructed to do so by your teacher.

6. Do NOT take any materials or chemicals outside of the laboratory.

7. Stay out of storage areas unless instructed to be there and supervised by your teacher.

Laboratory Cleanup

1. Turn off all burners, water, and gas, and disconnect all electrical devices.

2. Clean all pieces of equipment and return all materials to their proper places.

3. Dispose of chemicals and other materials as directed by your teacher. Place broken glass and solid substances in the proper containers. Never discard materials in the sink.

4. Clean your work area.

5. Wash your hands with soap and water thoroughly BEFORE removing your goggles.

Emergencies

1. Report any fire, electrical shock, glassware breakage, spill, or injury, no matter how small, to your teacher immediately. Follow his or her instructions.

2. If your clothing should catch fire, STOP, DROP, and ROLL. If possible, smother it with the fire blanket or get under a safety shower. NEVER RUN.

3. If a fire should occur, turn off all gas and leave the room according to established procedures.

4. In most instances, your teacher will clean up spills. Do NOT attempt to clean up spills unless you are given permission and instructions to do so.

5. If chemicals come into contact with your eyes or skin, notify your teacher immediately. Use the eyewash or flush your skin or eyes with large quantities of water.

6. The fire extinguisher and first-aid kit should only be used by your teacher unless it is an extreme emergency and you have been given permission.

7. If someone is injured or becomes ill, only a professional medical provider or someone certified in first aid should perform first-aid procedures.

EXTRA Try at Home Labs

From Your Kitchen, Junk Drawer, or Yard

1 Comparing Particles

▶ Real-World Question

How do the masses of an electron, neutron, and proton compare?

Possible Materials

- large, self-sealing plastic bags or clear 2-L bottles (6)
- liquid measuring cup
- marker
- large poster paper

▶ Procedure

1. Measure exactly 1 mL of water into a plastic bag. Seal it. Write *Electron* on it. Make another bag exactly the same.
2. Measure exactly 1,837 mL into another plastic bag. Seal it. Write *Proton* on it. Make another bag exactly the same.
3. Measure exactly 1,839 mL into another plastic bag. Seal it. Write *Neutron* on it. Make another bag exactly the same.
4. Draw a large helium atom on the paper. Put the four bags representing two neutrons and the two protons in the central nucleus. Put the two electron bags at a distance from the nucleus, representing their orbit around the nucleus.

▶ Conclude and Apply

1. Compare the mass of the proton and neutron.
2. Infer why nearly all the mass of an atom is located in its nucleus.

2 Microscopic Crystals

▶ Real-World Question

What do crystalline and non-crystalline solids look like under a magnifying lens?

Possible Materials

- salt or sugar
- pepper
- magnifying lens
- paper
- bowl
- spoon
- measuring cup

▶ Procedure

1. Pour 10 mL of salt into a bowl and grind the salt into small, powdery pieces with the back of the spoon.
2. Sprinkle a few grains of salt from the bowl onto a piece of paper and view the salt grains with the magnifying lens.
3. Clean out the bowl.
4. Pour 10 mL of pepper into the bowl and grind it into powder with the spoon.
5. Sprinkle a few grains of pepper from the bowl onto the paper and view the grains with the magnifying lens.

▶ Conclude and Apply

1. Compare the difference between the salt and pepper grains under the magnifying lens.
2. Describe what a crystal is.

Adult supervision required for all labs.

3 Good and Bad Apples

▶ Real-World Question
How can the chemical reaction that turns apples brown be stopped?

Possible Materials
- apple
- concentrated lemon juice
- orange juice
- vitamin C tablet (1000 mg)
- water
- cola
- bowls (5)
- measuring cup
- kitchen knife
- paper plates (6)
- black marker

▶ Procedure
1. Cut an apple into six equal slices.
2. Place one slice on a paper plate and label the plate *Untreated*.
3. Pour 100 mL of water into the first two bowls.
4. Dissolve a vitamin C tablet in the second bowl of water.
5. Pour 100 mL lemon juice, 100 mL of orange juice, and 100 mL of cola into the three remaining bowls.
6. Submerge an apple slice in each bowl for 10 min.
7. Label your other five plates *Water, Vitamin C Water, Lemon Juice, Orange Juice,* and *Cola.*
8. Take your apple wedges out of the bowls, place them on their correct plates. Observe the slices after one hour.

▶ Conclude and Apply
1. Describe the results of your experiment.
2. Infer why some apple slices did not turn brown after being submerged.

4 Research Race

▶ Real-World Question
How many secrets of the periodic table's elements can you find by research?

Possible Materials
- reference materials
- access to library

▶ Procedure
1. Get together with a team of your friends. Look at the Race Questions and divide them between you.
2. Try to get as many answers as you can in a certain amount of time.
3. Be sure to keep a record of each resource. You get a point for each correct answer. You also get a point for each properly listed book, magazine, or Web site that you list.

Race Questions:
- List colored compounds of transition metals.
- List uses of colored transition metal compounds.
- List elements that are dangerous to human health, especially heavy metals. Where are they found in society?
- List elements that are needed for human health. What food sources are each found in?
- List any other interesting information about elements that show up as you do your research.

▶ Conclude and Apply
1. Which resources did you find most helpful?
2. Name an interesting fact you found.

Computer Skills

People who study science rely on computers, like the one in **Figure 16,** to record and store data and to analyze results from investigations. Whether you work in a laboratory or just need to write a lab report with tables, good computer skills are a necessity.

Using the computer comes with responsibility. Issues of ownership, security, and privacy can arise. Remember, if you did not author the information you are using, you must provide a source for your information. Also, anything on a computer can be accessed by others. Do not put anything on the computer that you would not want everyone to know. To add more security to your work, use a password.

Use a Word Processing Program

A computer program that allows you to type your information, change it as many times as you need to, and then print it out is called a word processing program. Word processing programs also can be used to make tables.

Figure 16 A computer will make reports neater and more professional looking.

Learn the Skill To start your word processing program, a blank document, sometimes called "Document 1," appears on the screen. To begin, start typing. To create a new document, click the *New* button on the standard tool bar. These tips will help you format the document.

- The program will automatically move to the next line; press *Enter* if you wish to start a new paragraph.

- Symbols, called non-printing characters, can be hidden by clicking the *Show/Hide* button on your toolbar.

- To insert text, move the cursor to the point where you want the insertion to go, click on the mouse once, and type the text.

- To move several lines of text, select the text and click the *Cut* button on your toolbar. Then position your cursor in the location that you want to move the cut text and click *Paste.* If you move to the wrong place, click *Undo.*

- The spell check feature does not catch words that are misspelled to look like other words, like "cold" instead of "gold." Always reread your document to catch all spelling mistakes.

- To learn about other word processing methods, read the user's manual or click on the *Help* button.

- You can integrate databases, graphics, and spreadsheets into documents by copying from another program and pasting it into your document, or by using desktop publishing (DTP). DTP software allows you to put text and graphics together to finish your document with a professional look. This software varies in how it is used and its capabilities.

Use a Database

A collection of facts stored in a computer and sorted into different fields is called a database. A database can be reorganized in any way that suits your needs.

Learn the Skill A computer program that allows you to create your own database is a database management system (DBMS). It allows you to add, delete, or change information. Take time to get to know the features of your database software.

- Determine what facts you would like to include and research to collect your information.
- Determine how you want to organize the information.
- Follow the instructions for your particular DBMS to set up fields. Then enter each item of data in the appropriate field.
- Follow the instructions to sort the information in order of importance.
- Evaluate the information in your database, and add, delete, or change as necessary.

Use the Internet

The Internet is a global network of computers where information is stored and shared. To use the Internet, like the students in **Figure 17,** you need a modem to connect your computer to a phone line and an Internet Service Provider account.

Learn the Skill To access internet sites and information, use a "Web browser," which lets you view and explore pages on the World Wide Web. Each page is its own site, and each site has its own address, called a URL. Once you have found a Web browser, follow these steps for a search (this also is how you search a database).

Figure 17 The Internet allows you to search a global network for a variety of information.

- Be as specific as possible. If you know you want to research "gold," don't type in "elements." Keep narrowing your search until you find what you want.
- Web sites that end in *.com* are commercial Web sites; *.org, .edu,* and *.gov* are nonprofit, educational, or government Web sites.
- Electronic encyclopedias, almanacs, indexes, and catalogs will help locate and select relevant information.
- Develop a "home page" with relative ease. When developing a Web site, NEVER post pictures or disclose personal information such as location, names, or phone numbers. Your school or community usually can host your Web site. A basic understanding of HTML (hypertext mark-up language), the language of Web sites, is necessary. Software that creates HTML code is called authoring software, and can be downloaded free from many Web sites. This software allows text and pictures to be arranged as the software is writing the HTML code.

Technology Skill Handbook

Use a Spreadsheet

A spreadsheet, shown in **Figure 18,** can perform mathematical functions with any data arranged in columns and rows. By entering a simple equation into a cell, the program can perform operations in specific cells, rows, or columns.

Learn the Skill Each column (vertical) is assigned a letter, and each row (horizontal) is assigned a number. Each point where a row and column intersect is called a cell, and is labeled according to where it is located— Column A, Row 1 (A1).

- Decide how to organize the data, and enter it in the correct row or column.
- Spreadsheets can use standard formulas or formulas can be customized to calculate cells.
- To make a change, click on a cell to make it activate, and enter the edited data or formula.
- Spreadsheets also can display your results in graphs. Choose the style of graph that best represents the data.

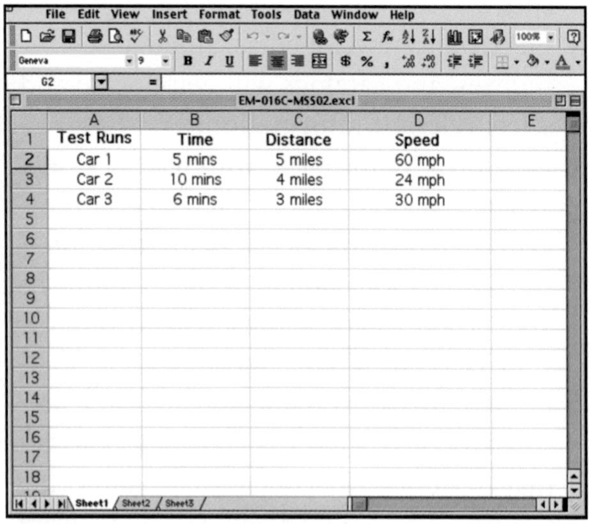

Figure 18 A spreadsheet allows you to perform mathematical operations on your data.

Use Graphics Software

Adding pictures, called graphics, to your documents is one way to make your documents more meaningful and exciting. This software adds, edits, and even constructs graphics. There is a variety of graphics software programs. The tools used for drawing can be a mouse, keyboard, or other specialized devices. Some graphics programs are simple. Others are complicated, called computer-aided design (CAD) software.

Learn the Skill It is important to have an understanding of the graphics software being used before starting. The better the software is understood, the better the results. The graphics can be placed in a word-processing document.

- Clip art can be found on a variety of internet sites, and on CDs. These images can be copied and pasted into your document.
- When beginning, try editing existing drawings, then work up to creating drawings.
- The images are made of tiny rectangles of color called pixels. Each pixel can be altered.
- Digital photography is another way to add images. The photographs in the memory of a digital camera can be downloaded into a computer, then edited and added to the document.
- Graphics software also can allow animation. The software allows drawings to have the appearance of movement by connecting basic drawings automatically. This is called in-betweening, or tweening.
- Remember to save often.

144 ◆ **K** STUDENT RESOURCES

Presentation Skills

Develop Multimedia Presentations

Most presentations are more dynamic if they include diagrams, photographs, videos, or sound recordings, like the one shown in **Figure 19.** A multimedia presentation involves using stereos, overhead projectors, televisions, computers, and more.

Learn the Skill Decide the main points of your presentation, and what types of media would best illustrate those points.

- Make sure you know how to use the equipment you are working with.
- Practice the presentation using the equipment several times.
- Enlist the help of a classmate to push play or turn lights out for you. Be sure to practice your presentation with him or her.
- If possible, set up all of the equipment ahead of time, and make sure everything is working properly.

Figure 19 These students are engaging the audience using a variety of tools.

Computer Presentations

There are many different interactive computer programs that you can use to enhance your presentation. Most computers have a compact disc (CD) drive that can play both CDs and digital video discs (DVDs). Also, there is hardware to connect a regular CD, DVD, or VCR. These tools will enhance your presentation.

Another method of using the computer to aid in your presentation is to develop a slide show using a computer program. This can allow movement of visuals at the presenter's pace, and can allow for visuals to build on one another.

Learn the Skill In order to create multimedia presentations on a computer, you need to have certain tools. These may include traditional graphic tools and drawing programs, animation programs, and authoring systems that tie everything together. Your computer will tell you which tools it supports. The most important step is to learn about the tools that you will be using.

- Often, color and strong images will convey a point better than words alone. Use the best methods available to convey your point.
- As with other presentations, practice many times.
- Practice your presentation with the tools you and any assistants will be using.
- Maintain eye contact with the audience. The purpose of using the computer is not to prompt the presenter, but to help the audience understand the points of the presentation.

Math Review

Use Fractions

A fraction compares a part to a whole. In the fraction $\frac{2}{3}$, the 2 represents the part and is the numerator. The 3 represents the whole and is the denominator.

Reduce Fractions To reduce a fraction, you must find the largest factor that is common to both the numerator and the denominator, the greatest common factor (GCF). Divide both numbers by the GCF. The fraction has then been reduced, or it is in its simplest form.

Example Twelve of the 20 chemicals in the science lab are in powder form. What fraction of the chemicals used in the lab are in powder form?

Step 1 Write the fraction.

$$\frac{\text{part}}{\text{whole}} = \frac{12}{20}$$

Step 2 To find the GCF of the numerator and denominator, list all of the factors of each number.

Factors of 12: 1, 2, 3, 4, 6, 12 (the numbers that divide evenly into 12)

Factors of 20: 1, 2, 4, 5, 10, 20 (the numbers that divide evenly into 20)

Step 3 List the common factors.

1, 2, 4.

Step 4 Choose the greatest factor in the list.

The GCF of 12 and 20 is 4.

Step 5 Divide the numerator and denominator by the GCF.

$$\frac{12 \div 4}{20 \div 4} = \frac{3}{5}$$

In the lab, $\frac{3}{5}$ of the chemicals are in powder form.

Practice Problem At an amusement park, 66 of 90 rides have a height restriction. What fraction of the rides, in its simplest form, has a height restriction?

Add and Subtract Fractions To add or subtract fractions with the same denominator, add or subtract the numerators and write the sum or difference over the denominator. After finding the sum or difference, find the simplest form for your fraction.

Example 1 In the forest outside your house, $\frac{1}{8}$ of the animals are rabbits, $\frac{3}{8}$ are squirrels, and the remainder are birds and insects. How many are mammals?

Step 1 Add the numerators.

$$\frac{1}{8} + \frac{3}{8} = \frac{(1+3)}{8} = \frac{4}{8}$$

Step 2 Find the GCF.

$$\frac{4}{8} \text{ (GCF, 4)}$$

Step 3 Divide the numerator and denominator by the GCF.

$$\frac{4}{4} = 1, \ \frac{8}{4} = 2$$

$\frac{1}{2}$ of the animals are mammals.

Example 2 If $\frac{7}{16}$ of the Earth is covered by freshwater, and $\frac{1}{16}$ of that is in glaciers, how much freshwater is not frozen?

Step 1 Subtract the numerators.

$$\frac{7}{16} - \frac{1}{16} = \frac{(7-1)}{16} = \frac{6}{16}$$

Step 2 Find the GCF.

$$\frac{6}{16} \text{ (GCF, 2)}$$

Step 3 Divide the numerator and denominator by the GCF.

$$\frac{6}{2} = 3, \ \frac{16}{2} = 8$$

$\frac{3}{8}$ of the freshwater is not frozen.

Practice Problem A bicycle rider is going 15 km/h for $\frac{4}{9}$ of his ride, 10 km/h for $\frac{2}{9}$ of his ride, and 8 km/h for the remainder of the ride. How much of his ride is he going over 8 km/h?

Unlike Denominators To add or subtract fractions with unlike denominators, first find the least common denominator (LCD). This is the smallest number that is a common multiple of both denominators. Rename each fraction with the LCD, and then add or subtract. Find the simplest form if necessary.

Example 1 A chemist makes a paste that is $\frac{1}{2}$ table salt (NaCl), $\frac{1}{3}$ sugar ($C_6H_{12}O_6$), and the rest water (H_2O). How much of the paste is a solid?

Step 1 Find the LCD of the fractions.

$\frac{1}{2} + \frac{1}{3}$ (LCD, 6)

Step 2 Rename each numerator and each denominator with the LCD.

$1 \times 3 = 3, \ 2 \times 3 = 6$

$1 \times 2 = 2, \ 3 \times 2 = 6$

Step 3 Add the numerators.

$\frac{3}{6} + \frac{2}{6} = \frac{(3+2)}{6} = \frac{5}{6}$

$\frac{5}{6}$ of the paste is a solid.

Example 2 The average precipitation in Grand Junction, CO, is $\frac{7}{10}$ inch in November, and $\frac{3}{5}$ inch in December. What is the total average precipitation?

Step 1 Find the LCD of the fractions.

$\frac{7}{10} + \frac{3}{5}$ (LCD, 10)

Step 2 Rename each numerator and each denominator with the LCD.

$7 \times 1 = 7, \ 10 \times 1 = 10$

$3 \times 2 = 6, \ 5 \times 2 = 10$

Step 3 Add the numerators.

$\frac{7}{10} + \frac{6}{10} = \frac{(7+6)}{10} = \frac{13}{10}$

$\frac{13}{10}$ inches total precipitation, or $1\frac{3}{10}$ inches.

Practice Problem On an electric bill, about $\frac{1}{8}$ of the energy is from solar energy and about $\frac{1}{10}$ is from wind power. How much of the total bill is from solar energy and wind power combined?

Example 3 In your body, $\frac{7}{10}$ of your muscle contractions are involuntary (cardiac and smooth muscle tissue). Smooth muscle makes $\frac{3}{15}$ of your muscle contractions. How many of your muscle contractions are made by cardiac muscle?

Step 1 Find the LCD of the fractions.

$\frac{7}{10} - \frac{3}{15}$ (LCD, 30)

Step 2 Rename each numerator and each denominator with the LCD.

$7 \times 3 = 21, \ 10 \times 3 = 30$

$3 \times 2 = 6, \ 15 \times 2 = 30$

Step 3 Subtract the numerators.

$\frac{21}{30} - \frac{6}{30} = \frac{(21-6)}{30} = \frac{15}{30}$

Step 4 Find the GCF.

$\frac{15}{30}$ (GCF, 15)

$\frac{1}{2}$

$\frac{1}{2}$ of all muscle contractions are cardiac muscle.

Example 4 Tony wants to make cookies that call for $\frac{3}{4}$ of a cup of flour, but he only has $\frac{1}{3}$ of a cup. How much more flour does he need?

Step 1 Find the LCD of the fractions.

$\frac{3}{4} - \frac{1}{3}$ (LCD, 12)

Step 2 Rename each numerator and each denominator with the LCD.

$3 \times 3 = 9, \ 4 \times 3 = 12$

$1 \times 4 = 4, \ 3 \times 4 = 12$

Step 3 Subtract the numerators.

$\frac{9}{12} - \frac{4}{12} = \frac{(9-4)}{12} = \frac{5}{12}$

$\frac{5}{12}$ of a cup of flour.

Practice Problem Using the information provided to you in Example 3 above, determine how many muscle contractions are voluntary (skeletal muscle).

Multiply Fractions To multiply with fractions, multiply the numerators and multiply the denominators. Find the simplest form if necessary.

Example Multiply $\frac{3}{5}$ by $\frac{1}{3}$.

Step 1 Multiply the numerators and denominators.
$$\frac{3}{5} \times \frac{1}{3} = \frac{(3 \times 1)}{(5 \times 3)} = \frac{3}{15}$$

Step 2 Find the GCF.
$$\frac{3}{15} \text{ (GCF, 3)}$$

Step 3 Divide the numerator and denominator by the GCF.
$$\frac{3}{3} = 1, \quad \frac{15}{3} = 5$$
$$\frac{1}{5}$$

$\frac{3}{5}$ multiplied by $\frac{1}{3}$ is $\frac{1}{5}$.

Practice Problem Multiply $\frac{3}{14}$ by $\frac{5}{16}$.

Find a Reciprocal Two numbers whose product is 1 are called multiplicative inverses, or reciprocals.

Example Find the reciprocal of $\frac{3}{8}$.

Step 1 Inverse the fraction by putting the denominator on top and the numerator on the bottom.
$$\frac{8}{3}$$

The reciprocal of $\frac{3}{8}$ is $\frac{8}{3}$.

Practice Problem Find the reciprocal of $\frac{4}{9}$.

Divide Fractions To divide one fraction by another fraction, multiply the dividend by the reciprocal of the divisor. Find the simplest form if necessary.

Example 1 Divide $\frac{1}{9}$ by $\frac{1}{3}$.

Step 1 Find the reciprocal of the divisor.
The reciprocal of $\frac{1}{3}$ is $\frac{3}{1}$.

Step 2 Multiply the dividend by the reciprocal of the divisor.
$$\frac{\frac{1}{9}}{\frac{1}{3}} = \frac{1}{9} \times \frac{3}{1} = \frac{(1 \times 3)}{(9 \times 1)} = \frac{3}{9}$$

Step 3 Find the GCF.
$$\frac{3}{9} \text{ (GCF, 3)}$$

Step 4 Divide the numerator and denominator by the GCF.
$$\frac{3}{3} = 1, \quad \frac{9}{3} = 3$$
$$\frac{1}{3}$$

$\frac{1}{9}$ divided by $\frac{1}{3}$ is $\frac{1}{3}$.

Example 2 Divide $\frac{3}{5}$ by $\frac{1}{4}$.

Step 1 Find the reciprocal of the divisor.
The reciprocal of $\frac{1}{4}$ is $\frac{4}{1}$.

Step 2 Multiply the dividend by the reciprocal of the divisor.
$$\frac{\frac{3}{5}}{\frac{1}{4}} = \frac{3}{5} \times \frac{4}{1} = \frac{(3 \times 4)}{(5 \times 1)} = \frac{12}{5}$$

$\frac{3}{5}$ divided by $\frac{1}{4}$ is $\frac{12}{5}$ or $2\frac{2}{5}$.

Practice Problem Divide $\frac{3}{11}$ by $\frac{7}{10}$.

Use Ratios

When you compare two numbers by division, you are using a ratio. Ratios can be written 3 to 5, 3:5, or $\frac{3}{5}$. Ratios, like fractions, also can be written in simplest form.

Ratios can represent probabilities, also called odds. This is a ratio that compares the number of ways a certain outcome occurs to the number of outcomes. For example, if you flip a coin 100 times, what are the odds that it will come up heads? There are two possible outcomes, heads or tails, so the odds of coming up heads are 50:100. Another way to say this is that 50 out of 100 times the coin will come up heads. In its simplest form, the ratio is 1:2.

Example 1 A chemical solution contains 40 g of salt and 64 g of baking soda. What is the ratio of salt to baking soda as a fraction in simplest form?

Step 1 Write the ratio as a fraction.
$$\frac{\text{salt}}{\text{baking soda}} = \frac{40}{64}$$

Step 2 Express the fraction in simplest form.
The GCF of 40 and 64 is 8.
$$\frac{40}{64} = \frac{40 \div 8}{64 \div 8} = \frac{5}{8}$$

The ratio of salt to baking soda in the sample is 5:8.

Example 2 Sean rolls a 6-sided die 6 times. What are the odds that the side with a 3 will show?

Step 1 Write the ratio as a fraction.
$$\frac{\text{number of sides with a 3}}{\text{number of sides}} = \frac{1}{6}$$

Step 2 Multiply by the number of attempts.
$$\frac{1}{6} \times 6 \text{ attempts} = \frac{6}{6} \text{ attempts} = 1 \text{ attempt}$$

1 attempt out of 6 will show a 3.

Practice Problem Two metal rods measure 100 cm and 144 cm in length. What is the ratio of their lengths in simplest form?

Use Decimals

A fraction with a denominator that is a power of ten can be written as a decimal. For example, 0.27 means $\frac{27}{100}$. The decimal point separates the ones place from the tenths place.

Any fraction can be written as a decimal using division. For example, the fraction $\frac{5}{8}$ can be written as a decimal by dividing 5 by 8. Written as a decimal, it is 0.625.

Add or Subtract Decimals When adding and subtracting decimals, line up the decimal points before carrying out the operation.

Example 1 Find the sum of 47.68 and 7.80.

Step 1 Line up the decimal places when you write the numbers.
```
  47.68
+  7.80
```

Step 2 Add the decimals.
```
  47.68
+  7.80
  55.48
```

The sum of 47.68 and 7.80 is 55.48.

Example 2 Find the difference of 42.17 and 15.85.

Step 1 Line up the decimal places when you write the number.
```
  42.17
- 15.85
```

Step 2 Subtract the decimals.
```
  42.17
- 15.85
  26.32
```

The difference of 42.17 and 15.85 is 26.32.

Practice Problem Find the sum of 1.245 and 3.842.

Multiply Decimals To multiply decimals, multiply the numbers like any other number, ignoring the decimal point. Count the decimal places in each factor. The product will have the same number of decimal places as the sum of the decimal places in the factors.

Example Multiply 2.4 by 5.9.

Step 1 Multiply the factors like two whole numbers.
$$24 \times 59 = 1416$$

Step 2 Find the sum of the number of decimal places in the factors. Each factor has one decimal place, for a sum of two decimal places.

Step 3 The product will have two decimal places.
14.16

The product of 2.4 and 5.9 is 14.16.

Practice Problem Multiply 4.6 by 2.2.

Divide Decimals When dividing decimals, change the divisor to a whole number. To do this, multiply both the divisor and the dividend by the same power of ten. Then place the decimal point in the quotient directly above the decimal point in the dividend. Then divide as you do with whole numbers.

Example Divide 8.84 by 3.4.

Step 1 Multiply both factors by 10.
$$3.4 \times 10 = 34, \ 8.84 \times 10 = 88.4$$

Step 2 Divide 88.4 by 34.

```
        2.6
  34)88.4
    −68
     204
    −204
       0
```

8.84 divided by 3.4 is 2.6.

Practice Problem Divide 75.6 by 3.6.

Use Proportions

An equation that shows that two ratios are equivalent is a proportion. The ratios $\frac{2}{4}$ and $\frac{5}{10}$ are equivalent, so they can be written as $\frac{2}{4} = \frac{5}{10}$. This equation is a proportion.

When two ratios form a proportion, the cross products are equal. To find the cross products in the proportion $\frac{2}{4} = \frac{5}{10}$, multiply the 2 and the 10, and the 4 and the 5. Therefore $2 \times 10 = 4 \times 5$, or $20 = 20$.

Because you know that both proportions are equal, you can use cross products to find a missing term in a proportion. This is known as solving the proportion.

Example The heights of a tree and a pole are proportional to the lengths of their shadows. The tree casts a shadow of 24 m when a 6-m pole casts a shadow of 4 m. What is the height of the tree?

Step 1 Write a proportion.
$$\frac{\text{height of tree}}{\text{height of pole}} = \frac{\text{length of tree's shadow}}{\text{length of pole's shadow}}$$

Step 2 Substitute the known values into the proportion. Let h represent the unknown value, the height of the tree.
$$\frac{h}{6} = \frac{24}{4}$$

Step 3 Find the cross products.
$$h \times 4 = 6 \times 24$$

Step 4 Simplify the equation.
$$4h = 144$$

Step 5 Divide each side by 4.
$$\frac{4h}{4} = \frac{144}{4}$$
$$h = 36$$

The height of the tree is 36 m.

Practice Problem The ratios of the weights of two objects on the Moon and on Earth are in proportion. A rock weighing 3 N on the Moon weighs 18 N on Earth. How much would a rock that weighs 5 N on the Moon weigh on Earth?

Use Percentages

The word *percent* means "out of one hundred." It is a ratio that compares a number to 100. Suppose you read that 77 percent of the Earth's surface is covered by water. That is the same as reading that the fraction of the Earth's surface covered by water is $\frac{77}{100}$. To express a fraction as a percent, first find the equivalent decimal for the fraction. Then, multiply the decimal by 100 and add the percent symbol.

Example Express $\frac{13}{20}$ as a percent.

Step 1 Find the equivalent decimal for the fraction.

$$\begin{array}{r} 0.65 \\ 20\overline{)13.00} \\ \underline{12\ 0} \\ 1\ 00 \\ \underline{1\ 00} \\ 0 \end{array}$$

Step 2 Rewrite the fraction $\frac{13}{20}$ as 0.65.

Step 3 Multiply 0.65 by 100 and add the % sign.
$0.65 \times 100 = 65 = 65\%$

So, $\frac{13}{20} = 65\%$.

This also can be solved as a proportion.

Example Express $\frac{13}{20}$ as a percent.

Step 1 Write a proportion.
$$\frac{13}{20} = \frac{x}{100}$$

Step 2 Find the cross products.
$1300 = 20x$

Step 3 Divide each side by 20.
$$\frac{1300}{20} = \frac{20x}{20}$$
$65\% = x$

Practice Problem In one year, 73 of 365 days were rainy in one city. What percent of the days in that city were rainy?

Solve One-Step Equations

A statement that two things are equal is an equation. For example, $A = B$ is an equation that states that A is equal to B.

An equation is solved when a variable is replaced with a value that makes both sides of the equation equal. To make both sides equal the inverse operation is used. Addition and subtraction are inverses, and multiplication and division are inverses.

Example 1 Solve the equation $x - 10 = 35$.

Step 1 Find the solution by adding 10 to each side of the equation.
$x - 10 = 35$
$x - 10 + 10 = 35 + 10$
$x = 45$

Step 2 Check the solution.
$x - 10 = 35$
$45 - 10 = 35$
$35 = 35$

Both sides of the equation are equal, so $x = 45$.

Example 2 In the formula $a = bc$, find the value of c if $a = 20$ and $b = 2$.

Step 1 Rearrange the formula so the unknown value is by itself on one side of the equation by dividing both sides by b.
$$a = bc$$
$$\frac{a}{b} = \frac{bc}{b}$$
$$\frac{a}{b} = c$$

Step 2 Replace the variables a and b with the values that are given.
$$\frac{a}{b} = c$$
$$\frac{20}{2} = c$$
$$10 = c$$

Step 3 Check the solution.
$$a = bc$$
$$20 = 2 \times 10$$
$$20 = 20$$

Both sides of the equation are equal, so $c = 10$ is the solution when $a = 20$ and $b = 2$.

Practice Problem In the formula $h = gd$, find the value of d if $g = 12.3$ and $h = 17.4$.

Use Statistics

The branch of mathematics that deals with collecting, analyzing, and presenting data is statistics. In statistics, there are three common ways to summarize data with a single number—the mean, the median, and the mode.

The **mean** of a set of data is the arithmetic average. It is found by adding the numbers in the data set and dividing by the number of items in the set.

The **median** is the middle number in a set of data when the data are arranged in numerical order. If there were an even number of data points, the median would be the mean of the two middle numbers.

The **mode** of a set of data is the number or item that appears most often.

Another number that often is used to describe a set of data is the range. The **range** is the difference between the largest number and the smallest number in a set of data.

A **frequency table** shows how many times each piece of data occurs, usually in a survey. **Table 2** below shows the results of a student survey on favorite color.

Table 2 Student Color Choice		
Color	**Tally**	**Frequency**
red	||||	4
blue	卌	5
black	||	2
green	|||	3
purple	卌 ||	7
yellow	卌 |	6

Based on the frequency table data, which color is the favorite?

Example The speeds (in m/s) for a race car during five different time trials are 39, 37, 44, 36, and 44.

To find the mean:

Step 1 Find the sum of the numbers.
$$39 + 37 + 44 + 36 + 44 = 200$$

Step 2 Divide the sum by the number of items, which is 5.
$$200 \div 5 = 40$$

The mean is 40 m/s.

To find the median:

Step 1 Arrange the measures from least to greatest.
36, 37, 39, 44, 44

Step 2 Determine the middle measure.
36, 37, 39, 44, 44

The median is 39 m/s.

To find the mode:

Step 1 Group the numbers that are the same together.
44, 44, 36, 37, 39

Step 2 Determine the number that occurs most in the set.
44, 44, 36, 37, 39

The mode is 44 m/s.

To find the range:

Step 1 Arrange the measures from largest to smallest.
44, 44, 39, 37, 36

Step 2 Determine the largest and smallest measures in the set.
44, 44, 39, 37, 36

Step 3 Find the difference between the largest and smallest measures.
$$44 - 36 = 8$$

The range is 8 m/s.

Practice Problem Find the mean, median, mode, and range for the data set 8, 4, 12, 8, 11, 14, 16.

Use Geometry

The branch of mathematics that deals with the measurement, properties, and relationships of points, lines, angles, surfaces, and solids is called geometry.

Perimeter The **perimeter** (P) is the distance around a geometric figure. To find the perimeter of a rectangle, add the length and width and multiply that sum by two, or $2(l + w)$. To find perimeters of irregular figures, add the length of the sides.

Example 1 Find the perimeter of a rectangle that is 3 m long and 5 m wide.

Step 1 You know that the perimeter is 2 times the sum of the width and length.
$$P = 2(3 \text{ m} + 5 \text{ m})$$

Step 2 Find the sum of the width and length.
$$P = 2(8 \text{ m})$$

Step 3 Multiply by 2.
$$P = 16 \text{ m}$$

The perimeter is 16 m.

Example 2 Find the perimeter of a shape with sides measuring 2 cm, 5 cm, 6 cm, 3 cm.

Step 1 You know that the perimeter is the sum of all the sides.
$$P = 2 + 5 + 6 + 3$$

Step 2 Find the sum of the sides.
$$P = 2 + 5 + 6 + 3$$
$$P = 16$$

The perimeter is 16 cm.

Practice Problem Find the perimeter of a rectangle with a length of 18 m and a width of 7 m.

Practice Problem Find the perimeter of a triangle measuring 1.6 cm by 2.4 cm by 2.4 cm.

Area of a Rectangle The **area** (A) is the number of square units needed to cover a surface. To find the area of a rectangle, multiply the length times the width, or $l \times w$. When finding area, the units also are multiplied. Area is given in square units.

Example Find the area of a rectangle with a length of 1 cm and a width of 10 cm.

Step 1 You know that the area is the length multiplied by the width.
$$A = (1 \text{ cm} \times 10 \text{ cm})$$

Step 2 Multiply the length by the width. Also multiply the units.
$$A = 10 \text{ cm}^2$$

The area is 10 cm^2.

Practice Problem Find the area of a square whose sides measure 4 m.

Area of a Triangle To find the area of a triangle, use the formula:

$$A = \frac{1}{2}(\text{base} \times \text{height})$$

The base of a triangle can be any of its sides. The height is the perpendicular distance from a base to the opposite endpoint, or vertex.

Example Find the area of a triangle with a base of 18 m and a height of 7 m.

Step 1 You know that the area is $\frac{1}{2}$ the base times the height.
$$A = \frac{1}{2}(18 \text{ m} \times 7 \text{ m})$$

Step 2 Multiply $\frac{1}{2}$ by the product of 18 $\times$ 7. Multiply the units.
$$A = \frac{1}{2}(126 \text{ m}^2)$$
$$A = 63 \text{ m}^2$$

The area is 63 m^2.

Practice Problem Find the area of a triangle with a base of 27 cm and a height of 17 cm.

Circumference of a Circle The **diameter** (d) of a circle is the distance across the circle through its center, and the **radius** (r) is the distance from the center to any point on the circle. The radius is half of the diameter. The distance around the circle is called the **circumference** (C). The formula for finding the circumference is:

$$C = 2\pi r \ \ or \ \ C = \pi d$$

The circumference divided by the diameter is always equal to 3.1415926... This nonterminating and nonrepeating number is represented by the Greek letter π (pi). An approximation often used for π is 3.14.

Example 1 Find the circumference of a circle with a radius of 3 m.

Step 1 You know the formula for the circumference is 2 times the radius times π.
$$C = 2\pi(3)$$

Step 2 Multiply 2 times the radius.
$$C = 6\pi$$

Step 3 Multiply by π.
$$C = 19 \ m$$

The circumference is 19 m.

Example 2 Find the circumference of a circle with a diameter of 24.0 cm.

Step 1 You know the formula for the circumference is the diameter times π.
$$C = \pi(24.0)$$

Step 2 Multiply the diameter by π.
$$C = 75.4 \ cm$$

The circumference is 75.4 cm.

Practice Problem Find the circumference of a circle with a radius of 19 cm.

Area of a Circle The formula for the area of a circle is:
$$A = \pi r^2$$

Example 1 Find the area of a circle with a radius of 4.0 cm.

Step 1 $A = \pi(4.0)^2$

Step 2 Find the square of the radius.
$$A = 16\pi$$

Step 3 Multiply the square of the radius by π.
$$A = 50 \ cm^2$$

The area of the circle is 50 cm^2.

Example 2 Find the area of a circle with a radius of 225 m.

Step 1 $A = \pi(225)^2$

Step 2 Find the square of the radius.
$$A = 50625\pi$$

Step 3 Multiply the square of the radius by π.
$$A = 158962.5$$

The area of the circle is 158,962 m^2.

Example 3 Find the area of a circle whose diameter is 20.0 mm.

Step 1 You know the formula for the area of a circle is the square of the radius times π, and that the radius is half of the diameter.
$$A = \pi\left(\frac{20.0}{2}\right)^2$$

Step 2 Find the radius.
$$A = \pi(10.0)^2$$

Step 3 Find the square of the radius.
$$A = 100\pi$$

Step 4 Multiply the square of the radius by π.
$$A = 314 \ mm^2$$

The area is 314 mm^2.

Practice Problem Find the area of a circle with a radius of 16 m.

Math Skill Handbook

Volume The measure of space occupied by a solid is the **volume** (V). To find the volume of a rectangular solid multiply the length times width times height, or $V = l \times w \times h$. It is measured in cubic units, such as cubic centimeters (cm^3).

Example Find the volume of a rectangular solid with a length of 2.0 m, a width of 4.0 m, and a height of 3.0 m.

Step 1 You know the formula for volume is the length times the width times the height.
$$V = 2.0\,m \times 4.0\,m \times 3.0\,m$$

Step 2 Multiply the length times the width times the height.
$$V = 24\,m^3$$

The volume is 24 m^3.

Practice Problem Find the volume of a rectangular solid that is 8 m long, 4 m wide, and 4 m high.

To find the volume of other solids, multiply the area of the base times the height.

Example 1 Find the volume of a solid that has a triangular base with a length of 8.0 m and a height of 7.0 m. The height of the entire solid is 15.0 m.

Step 1 You know that the base is a triangle, and the area of a triangle is $\frac{1}{2}$ the base times the height, and the volume is the area of the base times the height.
$$V = \left[\frac{1}{2}(b \times h)\right] \times 15$$

Step 2 Find the area of the base.
$$V = \left[\frac{1}{2}(8 \times 7)\right] \times 15$$
$$V = \left(\frac{1}{2} \times 56\right) \times 15$$

Step 3 Multiply the area of the base by the height of the solid.
$$V = 28 \times 15$$
$$V = 420\,m^3$$

The volume is 420 m^3.

Example 2 Find the volume of a cylinder that has a base with a radius of 12.0 cm, and a height of 21.0 cm.

Step 1 You know that the base is a circle, and the area of a circle is the square of the radius times π, and the volume is the area of the base times the height.
$$V = (\pi r^2) \times 21$$
$$V = (\pi 12^2) \times 21$$

Step 2 Find the area of the base.
$$V = 144\pi \times 21$$
$$V = 452 \times 21$$

Step 3 Multiply the area of the base by the height of the solid.
$$V = 9490\,cm^3$$

The volume is 9490 cm^3.

Example 3 Find the volume of a cylinder that has a diameter of 15 mm and a height of 4.8 mm.

Step 1 You know that the base is a circle with an area equal to the square of the radius times π. The radius is one-half the diameter. The volume is the area of the base times the height.
$$V = (\pi r^2) \times 4.8$$
$$V = \left[\pi\left(\frac{1}{2} \times 15\right)^2\right] \times 4.8$$
$$V = (\pi 7.5^2) \times 4.8$$

Step 2 Find the area of the base.
$$V = 56.25\pi \times 4.8$$
$$V = 176.63 \times 4.8$$

Step 3 Multiply the area of the base by the height of the solid.
$$V = 847.8$$

The volume is 847.8 mm^3.

Practice Problem Find the volume of a cylinder with a diameter of 7 cm in the base and a height of 16 cm.

Science Applications

Measure in SI

The metric system of measurement was developed in 1795. A modern form of the metric system, called the International System (SI), was adopted in 1960 and provides the standard measurements that all scientists around the world can understand.

The SI system is convenient because unit sizes vary by powers of 10. Prefixes are used to name units. Look at **Table 3** for some common SI prefixes and their meanings.

Table 3 Common SI Prefixes			
Prefix	**Symbol**	**Meaning**	
kilo-	k	1,000	thousand
hecto-	h	100	hundred
deka-	da	10	ten
deci-	d	0.1	tenth
centi-	c	0.01	hundredth
milli-	m	0.001	thousandth

Example How many grams equal one kilogram?

Step 1 Find the prefix *kilo* in **Table 3.**

Step 2 Using **Table 3,** determine the meaning of *kilo.* According to the table, it means 1,000. When the prefix *kilo* is added to a unit, it means that there are 1,000 of the units in a "*kilo*unit."

Step 3 Apply the prefix to the units in the question. The units in the question are grams. There are 1,000 grams in a kilogram.

Practice Problem Is a milligram larger or smaller than a gram? How many of the smaller units equal one larger unit? What fraction of the larger unit does one smaller unit represent?

Dimensional Analysis

Convert SI Units In science, quantities such as length, mass, and time sometimes are measured using different units. A process called dimensional analysis can be used to change one unit of measure to another. This process involves multiplying your starting quantity and units by one or more conversion factors. A conversion factor is a ratio equal to one and can be made from any two equal quantities with different units. If 1,000 mL equal 1 L then two ratios can be made.

$$\frac{1,000 \text{ mL}}{1 \text{ L}} = \frac{1 \text{ L}}{1,000 \text{ mL}} = 1$$

One can covert between units in the SI system by using the equivalents in **Table 3** to make conversion factors.

Example 1 How many cm are in 4 m?

Step 1 Write conversion factors for the units given. From **Table 3,** you know that 100 cm = 1 m. The conversion factors are

$$\frac{100 \text{ cm}}{1 \text{ m}} \quad and \quad \frac{1 \text{ m}}{100 \text{ cm}}$$

Step 2 Decide which conversion factor to use. Select the factor that has the units you are converting from (m) in the denominator and the units you are converting to (cm) in the numerator.

$$\frac{100 \text{ cm}}{1 \text{ m}}$$

Step 3 Multiply the starting quantity and units by the conversion factor. Cancel the starting units with the units in the denominator. There are 400 cm in 4 m.

$$4 \text{ m} \times \frac{100 \text{ cm}}{1 \text{ m}} = 400 \text{ cm}$$

Practice Problem How many milligrams are in one kilogram? (Hint: You will need to use two conversion factors from **Table 3.**)

Table 4 Unit System Equivalents

Type of Measurement	Equivalent
Length	1 in = 2.54 cm 1 yd = 0.91 m 1 mi = 1.61 km
Mass and Weight*	1 oz = 28.35 g 1 lb = 0.45 kg 1 ton (short) = 0.91 tonnes (metric tons) 1 lb = 4.45 N
Volume	$1 \text{ in}^3 = 16.39 \text{ cm}^3$ 1 qt = 0.95 L 1 gal = 3.78 L
Area	$1 \text{ in}^2 = 6.45 \text{ cm}^2$ $1 \text{ yd}^2 = 0.83 \text{ m}^2$ $1 \text{ mi}^2 = 2.59 \text{ km}^2$ 1 acre = 0.40 hectares
Temperature	$°C = \dfrac{(°F - 32)}{1.8}$ $K = °C + 273$

*Weight is measured in standard Earth gravity.

Convert Between Unit Systems **Table 4** gives a list of equivalents that can be used to convert between English and SI units.

Example If a meterstick has a length of 100 cm, how long is the meterstick in inches?

Step 1 Write the conversion factors for the units given. From **Table 4,** 1 in = 2.54 cm.

$$\frac{1 \text{ in}}{2.54 \text{ cm}} \quad and \quad \frac{2.54 \text{ cm}}{1 \text{ in}}$$

Step 2 Determine which conversion factor to use. You are converting from cm to in. Use the conversion factor with cm on the bottom.

$$\frac{1 \text{ in}}{2.54 \text{ cm}}$$

Step 3 Multiply the starting quantity and units by the conversion factor. Cancel the starting units with the units in the denominator. Round your answer based on the number of significant figures in the conversion factor.

$$100 \text{ cm} \times \frac{1 \text{ in}}{2.54 \text{ cm}} = 39.37 \text{ in}$$

The meterstick is 39.4 in long.

Practice Problem A book has a mass of 5 lbs. What is the mass of the book in kg?

Practice Problem Use the equivalent for in and cm (1 in = 2.54 cm) to show how $1 \text{ in}^3 = 16.39 \text{ cm}^3$.

MATH SKILL HANDBOOK **K** ◆ **157**

Precision and Significant Digits

When you make a measurement, the value you record depends on the precision of the measuring instrument. This precision is represented by the number of significant digits recorded in the measurement. When counting the number of significant digits, all digits are counted except zeros at the end of a number with no decimal point such as 2,050, and zeros at the beginning of a decimal such as 0.03020. When adding or subtracting numbers with different precision, round the answer to the smallest number of decimal places of any number in the sum or difference. When multiplying or dividing, the answer is rounded to the smallest number of significant digits of any number being multiplied or divided.

Example The lengths 5.28 and 5.2 are measured in meters. Find the sum of these lengths and record your answer using the correct number of significant digits.

Step 1 Find the sum.

$$
\begin{array}{ll}
5.28 \text{ m} & \text{2 digits after the decimal} \\
+\ 5.2\ \text{ m} & \text{1 digit after the decimal} \\
\hline
10.48 \text{ m} &
\end{array}
$$

Step 2 Round to one digit after the decimal because the least number of digits after the decimal of the numbers being added is 1.

The sum is 10.5 m.

Practice Problem How many significant digits are in the measurement 7,071,301 m? How many significant digits are in the measurement 0.003010 g?

Practice Problem Multiply 5.28 and 5.2 using the rule for multiplying and dividing. Record the answer using the correct number of significant digits.

Scientific Notation

Many times numbers used in science are very small or very large. Because these numbers are difficult to work with scientists use scientific notation. To write numbers in scientific notation, move the decimal point until only one non-zero digit remains on the left. Then count the number of places you moved the decimal point and use that number as a power of ten. For example, the average distance from the Sun to Mars is 227,800,000,000 m. In scientific notation, this distance is 2.278×10^{11} m. Because you moved the decimal point to the left, the number is a positive power of ten.

The mass of an electron is about 0.000 000 000 000 000 000 000 000 000 000 911 kg. Expressed in scientific notation, this mass is 9.11×10^{-31} kg. Because the decimal point was moved to the right, the number is a negative power of ten.

Example Earth is 149,600,000 km from the Sun. Express this in scientific notation.

Step 1 Move the decimal point until one non-zero digit remains on the left.
1.496 000 00

Step 2 Count the number of decimal places you have moved. In this case, eight.

Step 3 Show that number as a power of ten, 10^8.

The Earth is 1.496×10^8 km from the Sun.

Practice Problem How many significant digits are in 149,600,000 km? How many significant digits are in 1.496×10^8 km?

Practice Problem Parts used in a high performance car must be measured to 7×10^{-6} m. Express this number as a decimal.

Practice Problem A CD is spinning at 539 revolutions per minute. Express this number in scientific notation.

Make and Use Graphs

Data in tables can be displayed in a graph—a visual representation of data. Common graph types include line graphs, bar graphs, and circle graphs.

Line Graph A line graph shows a relationship between two variables that change continuously. The independent variable is changed and is plotted on the *x*-axis. The dependent variable is observed, and is plotted on the *y*-axis.

Example Draw a line graph of the data below from a cyclist in a long-distance race.

Table 5 Bicycle Race Data	
Time (h)	**Distance (km)**
0	0
1	8
2	16
3	24
4	32
5	40

Step 1 Determine the *x*-axis and *y*-axis variables. Time varies independently of distance and is plotted on the *x*-axis. Distance is dependent on time and is plotted on the *y*-axis.

Step 2 Determine the scale of each axis. The *x*-axis data ranges from 0 to 5. The *y*-axis data ranges from 0 to 40.

Step 3 Using graph paper, draw and label the axes. Include units in the labels.

Step 4 Draw a point at the intersection of the time value on the *x*-axis and corresponding distance value on the *y*-axis. Connect the points and label the graph with a title, as shown in **Figure 20.**

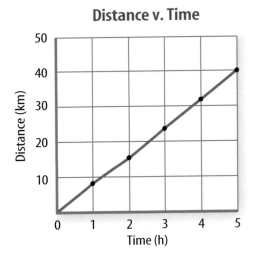

Distance v. Time

Figure 20 This line graph shows the relationship between distance and time during a bicycle ride.

Practice Problem A puppy's shoulder height is measured during the first year of her life. The following measurements were collected: (3 mo, 52 cm), (6 mo, 72 cm), (9 mo, 83 cm), (12 mo, 86 cm). Graph this data.

Find a Slope The slope of a straight line is the ratio of the vertical change, rise, to the horizontal change, run.

$$\text{Slope} = \frac{\text{vertical change (rise)}}{\text{horizontal change (run)}} = \frac{\text{change in } y}{\text{change in } x}$$

Example Find the slope of the graph in **Figure 20.**

Step 1 You know that the slope is the change in *y* divided by the change in *x*.
$$\text{Slope} = \frac{\text{change in } y}{\text{change in } x}$$

Step 2 Determine the data points you will be using. For a straight line, choose the two sets of points that are the farthest apart.
$$\text{Slope} = \frac{(40-0) \text{ km}}{(5-0) \text{ hr}}$$

Step 3 Find the change in *y* and *x*.
$$\text{Slope} = \frac{40 \text{ km}}{5 \text{ h}}$$

Step 4 Divide the change in *y* by the change in *x*.
$$\text{Slope} = \frac{8 \text{ km}}{\text{h}}$$

The slope of the graph is 8 km/h.

Bar Graph To compare data that does not change continuously you might choose a bar graph. A bar graph uses bars to show the relationships between variables. The *x*-axis variable is divided into parts. The parts can be numbers such as years, or a category such as a type of animal. The *y*-axis is a number and increases continuously along the axis.

Example A recycling center collects 4.0 kg of aluminum on Monday, 1.0 kg on Wednesday, and 2.0 kg on Friday. Create a bar graph of this data.

Step 1 Select the *x*-axis and *y*-axis variables. The measured numbers (the masses of aluminum) should be placed on the *y*-axis. The variable divided into parts (collection days) is placed on the *x*-axis.

Step 2 Create a graph grid like you would for a line graph. Include labels and units.

Step 3 For each measured number, draw a vertical bar above the *x*-axis value up to the *y*-axis value. For the first data point, draw a vertical bar above Monday up to 4.0 kg.

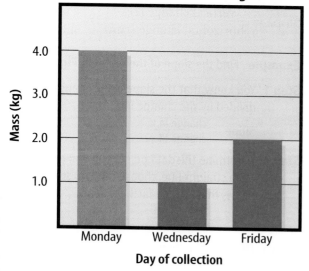

Aluminum Collected During Week

Practice Problem Draw a bar graph of the gases in air: 78% nitrogen, 21% oxygen, 1% other gases.

Circle Graph To display data as parts of a whole, you might use a circle graph. A circle graph is a circle divided into sections that represent the relative size of each piece of data. The entire circle represents 100%, half represents 50%, and so on.

Example Air is made up of 78% nitrogen, 21% oxygen, and 1% other gases. Display the composition of air in a circle graph.

Step 1 Multiply each percent by 360° and divide by 100 to find the angle of each section in the circle.

$$78\% \times \frac{360°}{100} = 280.8°$$

$$21\% \times \frac{360°}{100} = 75.6°$$

$$1\% \times \frac{360°}{100} = 3.6°$$

Step 2 Use a compass to draw a circle and to mark the center of the circle. Draw a straight line from the center to the edge of the circle.

Step 3 Use a protractor and the angles you calculated to divide the circle into parts. Place the center of the protractor over the center of the circle and line the base of the protractor over the straight line.

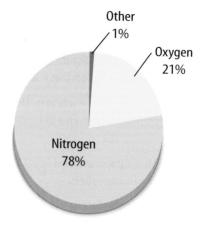

Practice Problem Draw a circle graph to represent the amount of aluminum collected during the week shown in the bar graph to the left.

Physical Science Reference Tables

Standard Units

Symbol	Name	Quantity
m	meter	length
kg	kilogram	mass
Pa	pascal	pressure
K	kelvin	temperature
mol	mole	amount of a substance
J	joule	energy, work, quantity of heat
s	second	time
C	coulomb	electric charge
V	volt	electric potential
A	ampere	electric current
Ω	ohm	resistance

Wavelengths of Light in a Vacuum

Violet	$4.0 - 4.2 \times 10^{-7}$ m
Blue	$4.2 - 4.9 \times 10^{-7}$ m
Green	$4.9 - 5.7 \times 10^{-7}$ m
Yellow	$5.7 - 5.9 \times 10^{-7}$ m
Orange	$5.9 - 6.5 \times 10^{-7}$ m
Red	$6.5 - 7.0 \times 10^{-7}$ m

Physical Constants and Conversion Factors

Acceleration due to gravity	g	9.8 m/s/s or m/s^2
Avogadro's Number	N$_A$	6.02×10^{23} particles per mole
Electron charge	e	1.6×10^{-19} C
Electron rest mass	m$_e$	9.11×10^{-31} kg
Gravitation constant	G	6.67×10^{-11} N $\times$ m^2/kg^2
Mass-energy relationship		1 u (amu) $= 9.3 \times 10^2$ MeV
Speed of light in a vacuum	c	3.00×108 m/s
Speed of sound at STP		331 m/s
Standard Pressure		1 atmosphere
		101.3 kPa
		760 Torr or mmHg
		14.7 lb/in.^{2}

The Index of Refraction for Common Substances

($\lambda = 5.9 \times 10^{-7}$ m)

Air	1.00
Alcohol	1.36
Canada Balsam	1.53
Corn Oil	1.47
Diamond	2.42
Glass, Crown	1.52
Glass, Flint	1.61
Glycerol	1.47
Lucite	1.50
Quartz, Fused	1.46
Water	1.33

Heat Constants

	Specific Heat (average) (kJ/kg $\times$ °C) (J/g $\times$ °C)	Melting Point (°C)	Boiling Point (°C)	Heat of Fusion (kJ/kg) (J/g)	Heat of Vaporization (kJ/kg) (J/g)
Alcohol (ethyl)	2.43 (liq.)	−117	79	109	855
Aluminum	0.90 (sol.)	660	2467	396	10500
Ammonia	4.71 (liq.)	−78	−33	332	1370
Copper	0.39 (sol.)	1083	2567	205	4790
Iron	0.45 (sol.)	1535	2750	267	6290
Lead	0.13 (sol.)	328	1740	25	866
Mercury	0.14 (liq.)	−39	357	11	295
Platinum	0.13 (sol.)	1772	3827	101	229
Silver	0.24 (sol.)	962	2212	105	2370
Tungsten	0.13 (sol.)	3410	5660	192	4350
Water (solid)	2.05 (sol.)	0	–	334	–
Water (liquid)	4.18 (liq.)	–	100	–	–
Water (vapor)	2.01 (gas)	–	–	–	2260
Zinc	0.39 (sol.)	420	907	113	1770

PERIODIC TABLE OF THE ELEMENTS

Columns of elements are called groups. Elements in the same group have similar chemical properties.

Gas
Liquid
Solid
Synthetic

Element — Hydrogen
Atomic number — 1
Symbol — **H**
Atomic mass — 1.008

State of matter

The first three symbols tell you the state of matter of the element at room temperature. The fourth symbol identifies elements that are not present in significant amounts on Earth. Useful amounts are made synthetically.

1

1 | Hydrogen 1 **H** 1.008

2

2 | Lithium 3 **Li** 6.941 | Beryllium 4 **Be** 9.012

3 | Sodium 11 **Na** 22.990 | Magnesium 12 **Mg** 24.305

3 **4** **5** **6** **7** **8** **9**

4 | Potassium 19 **K** 39.098 | Calcium 20 **Ca** 40.078 | Scandium 21 **Sc** 44.956 | Titanium 22 **Ti** 47.867 | Vanadium 23 **V** 50.942 | Chromium 24 **Cr** 51.996 | Manganese 25 **Mn** 54.938 | Iron 26 **Fe** 55.845 | Cobalt 27 **Co** 58.933

5 | Rubidium 37 **Rb** 85.468 | Strontium 38 **Sr** 87.62 | Yttrium 39 **Y** 88.906 | Zirconium 40 **Zr** 91.224 | Niobium 41 **Nb** 92.906 | Molybdenum 42 **Mo** 95.94 | Technetium 43 **Tc** (98) | Ruthenium 44 **Ru** 101.07 | Rhodium 45 **Rh** 102.906

6 | Cesium 55 **Cs** 132.905 | Barium 56 **Ba** 137.327 | Lanthanum 57 **La** 138.906 | Hafnium 72 **Hf** 178.49 | Tantalum 73 **Ta** 180.948 | Tungsten 74 **W** 183.84 | Rhenium 75 **Re** 186.207 | Osmium 76 **Os** 190.23 | Iridium 77 **Ir** 192.217

7 | Francium 87 **Fr** (223) | Radium 88 **Ra** (226) | Actinium 89 **Ac** (227) | Rutherfordium 104 **Rf** (261) | Dubnium 105 **Db** (262) | Seaborgium 106 **Sg** (266) | Bohrium 107 **Bh** (264) | Hassium 108 **Hs** (277) | Meitnerium 109 **Mt** (268)

The number in parentheses is the mass number of the longest-lived isotope for that element.

Rows of elements are called periods. Atomic number increases across a period.

The arrow shows where these elements would fit into the periodic table. They are moved to the bottom of the table to save space.

Lanthanide series | Cerium 58 **Ce** 140.116 | Praseodymium 59 **Pr** 140.908 | Neodymium 60 **Nd** 144.24 | Promethium 61 **Pm** (145) | Samarium 62 **Sm** 150.36

Actinide series | Thorium 90 **Th** 232.038 | Protactinium 91 **Pa** 231.036 | Uranium 92 **U** 238.029 | Neptunium 93 **Np** (237) | Plutonium 94 **Pu** (244)

Metal

Metalloid

Nonmetal

The color of an element's block tells you if the element is a metal, nonmetal, or metalloid.

Science Online
Visit bookk.msscience.com for updates to the periodic table.

18

Helium
2
He
4.003

13 **14** **15** **16** **17**

| Boron 5 **B** 10.811 | Carbon 6 **C** 12.011 | Nitrogen 7 **N** 14.007 | Oxygen 8 **O** 15.999 | Fluorine 9 **F** 18.998 | Neon 10 **Ne** 20.180 |

| Aluminum 13 **Al** 26.982 | Silicon 14 **Si** 28.086 | Phosphorus 15 **P** 30.974 | Sulfur 16 **S** 32.065 | Chlorine 17 **Cl** 35.453 | Argon 18 **Ar** 39.948 |

10 **11** **12**

| Nickel 28 **Ni** 58.693 | Copper 29 **Cu** 63.546 | Zinc 30 **Zn** 65.409 | Gallium 31 **Ga** 69.723 | Germanium 32 **Ge** 72.64 | Arsenic 33 **As** 74.922 | Selenium 34 **Se** 78.96 | Bromine 35 **Br** 79.904 | Krypton 36 **Kr** 83.798 |

| Palladium 46 **Pd** 106.42 | Silver 47 **Ag** 107.868 | Cadmium 48 **Cd** 112.411 | Indium 49 **In** 114.818 | Tin 50 **Sn** 118.710 | Antimony 51 **Sb** 121.760 | Tellurium 52 **Te** 127.60 | Iodine 53 **I** 126.904 | Xenon 54 **Xe** 131.293 |

| Platinum 78 **Pt** 195.078 | Gold 79 **Au** 196.967 | Mercury 80 **Hg** 200.59 | Thallium 81 **Tl** 204.383 | Lead 82 **Pb** 207.2 | Bismuth 83 **Bi** 208.980 | Polonium 84 **Po** (209) | Astatine 85 **At** (210) | Radon 86 **Rn** (222) |

| Darmstadtium 110 **Ds** (281) | Unununium * 111 **Uuu** (272) | Ununbium * 112 **Uub** (285) | | Ununquadium * 114 **Uuq** (289) | | ✶✶ 116 | | ✶✶ 118 |

* The names and symbols for elements 111–114 are temporary. Final names will be selected when the elements' discoveries are verified.
✶✶ Elements 116 and 118 were thought to have been created. The claim was retracted because the experimental results could not be repeated.

| Europium 63 **Eu** 151.964 | Gadolinium 64 **Gd** 157.25 | Terbium 65 **Tb** 158.925 | Dysprosium 66 **Dy** 162.500 | Holmium 67 **Ho** 164.930 | Erbium 68 **Er** 167.259 | Thulium 69 **Tm** 168.934 | Ytterbium 70 **Yb** 173.04 | Lutetium 71 **Lu** 174.967 |

| Americium 95 **Am** (243) | Curium 96 **Cm** (247) | Berkelium 97 **Bk** (247) | Californium 98 **Cf** (251) | Einsteinium 99 **Es** (252) | Fermium 100 **Fm** (257) | Mendelevium 101 **Md** (258) | Nobelium 102 **No** (259) | Lawrencium 103 **Lr** (262) |

Reference Handbooks

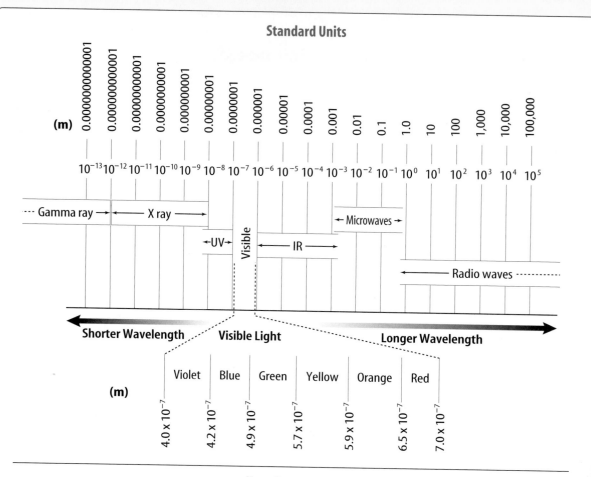

Standard Units

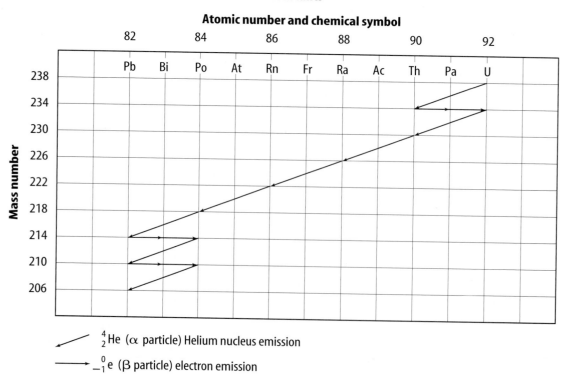

Heat Constants

Atomic number and chemical symbol

$\underset{2}{\overset{4}{}}$He ($\alpha$ particle) Helium nucleus emission

$\underset{-1}{\overset{0}{}}$e ($\beta$ particle) electron emission

Glossary/Glosario

Cómo usar el glosario en español:
1. Busca el término en inglés que desees encontrar.
2. El término en español, junto con la definición, se encuentran en la columna de la derecha.

Pronunciation Key

Use the following key to help you sound out words in the glossary.

a.............back (BAK)	ew.............food (FEWD)
ay.............day (DAY)	yoo...........pure (PYOOR)
ah.............father (FAH thur)	yew...........few (FYEW)
ow...........flower (FLOW ur)	uh.............comma (CAH muh)
ar.............car (CAR)	u (+ con).......rub (RUB)
e.............less (LES)	sh.............shelf (SHELF)
ee.............leaf (LEEF)	ch.............nature (NAY chur)
ih.............trip (TRIHP)	g.............gift (GIHFT)
i (i + con + e)..idea (i DEE uh)	j.............gem (JEM)
oh.............go (GOH)	ing.............sing (SING)
aw...........soft (SAWFT)	zh.............vision (VIH zhun)
or.............orbit (OR buht)	k.............cake (KAYK)
oy.............coin (COYN)	s.............seed, cent (SEED, SENT)
oo.............foot (FOOT)	z.............zone, raise (ZOHN, RAYZ)

English **Español**

actinide: the second series of inner transition elements which goes from thorium to lawrencium. (p. 114)

alkali metals: elements in group 1 of the periodic table. (p. 105)

alkaline earth metals: elements in group 2 of the periodic table. (p. 106)

Archimedes' (ar kuh MEE deez) principle: states that the buoyant force on an object is equal to the weight of the fluid displaced by the object. (p. 59)

atomic mass: average mass of an atom of an element; its unit of measure is the atomic mass unit (u), which is 1/12 the mass of a carbon-12 atom. (p. 22)

atomic number: number of protons in the nucleus of each atom of a given element; is the top number in the periodic table. (p. 21)

actínido: la segunda serie de los elementos de transición interna que abarca desde el torio hasta el laurencio. (p. 114)

metales alcalinos: elementos en el grupo 1 de la tabla periódica. (p. 105)

metales alcalinotérreos: elementos en el grupo 2 de la tabla periódica. (p. 106)

principio de Arquímedes: establece que la fuerza de empuje ejercida sobre un objeto es igual al peso del fluido desplazado por dicho objeto. (p. 59)

masa atómica: masa promedio de un átomo de un elemento; su unidad de medida es la unidad de masa atómica (u), la cual es 1/12 de la masa de un átomo de carbono 12. (p. 22)

número atómico: número de protones en el núcleo de cada átomo de un determinado elemento; es el número que se encuentra en la parte superior en la tabla periódica. (p. 21)

B

buoyant force: upward force exerted on an object immersed in a fluid. (p. 58)

fuerza de empuje: fuerza ascendente ejercida sobre un objeto inmerso en un fluido. (p. 58)

Glossary/Glosario

GLOSSARY/GLOSARIO **K** ◆ **165**

C

catalyst: substance that can make something happen faster but is not changed itself. (p. 113)

chemical change: change in which the composition of a substance changes. (p. 80)

chemical property: characteristic that cannot be observed without altering the sample. (p. 76)

compound: a substance produced when elements combine and whose properties are different from each of the elements in it. (p. 25)

condensation: the process of changing from a gas to a liquid. (pp. 51, 79)

catalizador: sustancia que puede hacer que algo suceda más rápidamente sin cambiar ella misma. (p. 113)

cambio químico: cambio en el cual la composición de una sustancia es modificada. (p. 80)

propiedad química: característica que no puede ser observada sin alterar la muestra. (p. 76)

compuesto: sustancia producida por la combinación de elementos y cuyas propiedades son diferentes de las de cada uno de los elementos. (p. 25)

condensación: el proceso de cambio de gas a líquido. (pp. 51, 79)

D

density: mass of an object divided by its volume. (p. 59)

deposition: the process by which a gas changes into a solid. (p. 79)

densidad: masa de un objeto dividida por su volumen. (p. 59)

deposición: el proceso mediante el cual un gas pasa a ser sólido. (p. 79)

E

electron: negatively-charged particle that exists in an electron cloud formation around an atom's nucleus. (p. 11)

electron cloud: region surrounding the nucleus of an atom, where electrons are most likely to be found. (p. 17)

element: substance that cannot be broken down into simpler substances. (p. 9)

electrón: partícula con carga negativa que existe en una nube de electrones alrededor del núcleo del átomo. (p. 11)

nube de electrones: región que rodea el núcleo de un átomo, en donde los electrones se encuentran con mayor probabilidad. (p. 17)

elemento: sustancia que no se puede descomponer en sustancias más simples. (p. 9)

F

freezing: change of matter from a liquid state to a solid state. (p. 49)

congelación: cambio de la materia de estado líquido a sólido. (p. 49)

G

gas: matter that does not have a definite shape or volume; has particles that move at high speeds in all directions. (p. 44)

group: family of elements in the periodic table that have similar physical or chemical properties. (p. 99)

gas: materia que no tiene ni forma ni volumen definidos; tiene partículas que se mueven a altas velocidades y en todas las direcciones. (p. 44)

grupo: familia de elementos en la tabla periódica que tienen propiedades físicas o químicas similares. (p. 99)

H

halogen: elements in group 17 of the periodic table. (p. 110)

heat: movement of thermal energy from a substance at a higher temperature to a substance at a lower temperature. (p. 46)

halógenos: elementos en el grupo 17 de la tabla periódica. (p. 110)

calor: movimiento de energía térmica de una sustancia que se encuentra a una alta temperatura hacia una sustancia a una baja temperatura. (p. 46)

I

isotopes (I suh tohps): two or more atoms of the same element that have different numbers of neutrons in their nuclei. (p. 21)

isótopos: dos o más átomos del mismo elemento que tienen diferente número de neutrones en sus núcleos. (p. 21)

L

lanthanide: the first series of inner transition elements which goes from cerium to lutetium. (p. 114)

law of conservation of mass: states that mass is neither created nor destroyed—and as a result the mass of the substances before a physical or chemical change is equal to the mass of the substances present after the change. (p. 87)

liquid: matter with a definite volume but no definite shape that can flow from one place to another. (p. 42)

lantánidos: la primera serie de los elementos de transición interna que va desde el cerio hasta el lutecio. (p. 114)

ley de la conservación de masas: establece que la masa no puede ser creada ni destruida; como resultado, la masa de una sustancia antes de un cambio físico o químico es igual a la masa presente de la sustancia después del cambio. (p. 87)

líquido: materia con volumen definido pero no con forma definida que puede fluir de un sitio a otro. (p. 42)

M

mass number: sum of the number of protons and neutrons in the nucleus of an atom. (p. 21)

matter: anything that takes up space and has mass. (p. 40)

melting: change of matter from a solid state to a liquid state. (p. 47)

metal: element that is malleable, ductile, a good conductor of electricity, and generally has a shiny or metallic luster. (pp. 22, 102)

metalloid (MEH tuh loyd): element that shares some properties with both metals and nonmetals. (pp. 23, 102)

mixture: a combination of compounds and elements that has not formed a new substance and whose proportions can be changed without changing the mixture's identity. (p. 27)

número de masa: suma del número de protones y neutrones en el núcleo de un átomo. (p. 21)

materia: cualquier cosa que ocupe espacio y tenga masa. (p. 40)

fusión: cambio de la materia de estado sólido a líquido. (p. 47)

metal: elemento maleable, dúctil, buen conductor de electricidad y generalmente con un lustre brillante o metálico. (pp. 22, 102)

metaloide: elemento que comparte algunas propiedades de los metales y de los no metales. (pp. 23, 102)

mezcla: combinación de compuestos y elementos sin llegar a formar una nueva sustancia y cuyas proporciones pueden cambiar sin que se modifique la identidad de la mezcla. (p. 27)

Glossary/Glosario

N

neutron (NEW trahn): electrically-neutral particle that has the same mass as a proton and is found in an atom's nucleus. (p. 15)

noble gases: elements in group 18 of the periodic table. (p. 110)

nonmetal: element that is usually a gas or brittle solid at room temperature and is a poor conductor of heat and electricity. (pp. 23, 102)

neutrón: partícula eléctricamente neutra que tiene la misma masa que un protón y se encuentra en el núcleo de un átomo. (p. 15)

gases inertes: elementos en el grupo 18 de la tabla periódica. (p. 110)

no metal: elemento que por lo general es un gas o un sólido frágil a temperatura ambiente y mal conductor de calor y electricidad. (pp. 23, 102)

P

Pascal's principle: states that when a force is applied to a confined fluid, an increase in pressure is transmitted equally to all parts of the fluid. (p. 60)

period: horizontal row of elements in the periodic table whose properties change gradually and predictably. (p. 99)

physical change: change in which the form or appearance of matter changes, but not its composition. (p. 78)

physical property: characteristic that can be observed, using the five senses, without changing or trying to change the composition of a substance. (p. 72)

pressure: force exerted on a surface divided by the total area over which the force is exerted. (p. 54)

proton: positively-charged particle in the nucleus of an atom. (p. 14)

principio de Pascal: establece que cuando se ejerce una fuerza sobre un fluido encerrado, se transmite un incremento de presión uniforme a todas las partes del fluido. (p. 60)

período: fila horizontal de elementos en la tabla periódica cuyas propiedades cambian gradualmente y en forma predecible. (p. 99)

cambio físico: cambio en el cual varía la forma o apariencia de la materia pero no su composición. (p. 78)

propiedad física: característica que puede ser observada usando los cinco sentidos sin cambiar o tratar de cambiar la composición de una sustancia. (p. 72)

presión: fuerza ejercida sobre una superficie dividida por el área total sobre la cual se ejerce dicha fuerza. (p. 54)

protón: partícula con carga positiva en el núcleo de un átomo. (p. 14)

R

representative elements: elements in groups 1 and 2 and 13–18 in the periodic table that include metals, metalloids, and nonmetals. (p. 99)

elementos representativos: elementos en los grupos 1 y 2 y 13-18 en la tabla periódica; incluyen metales, metaloides y no metales. (p. 99)

S

semiconductor: element that does not conduct electricity as well as a metal but conducts it better than a nonmetal. (p. 107)

solid: matter with a definite shape and volume; has tightly packed particles that move mainly by vibrating. (p. 41)

sublimation: the process by which a solid changes directly into a gas. (p. 79)

semiconductor: elemento que no conduce electricidad tan bien como un metal pero que la conduce mejor que un no metal. (p. 107)

sólido: materia con forma y volumen definidos; tiene partículas fuertemente compactadas que se mueven principalmente por vibración. (p. 41)

sublimación: proceso mediante el cual un sólido se convierte directamente en gas. (p. 79)

substance: matter that has the same composition and properties throughout. (p. 25)

surface tension: the uneven forces acting on the particles on the surface of a liquid. (p. 43)

sustancia: materia que tiene la misma composición y propiedades en cada una de sus partes. (p. 25)

tensión superficial: fuerzas desiguales que actúan sobre las partículas que se encuentran en la superficie de un líquido. (p. 43)

temperature: measure of the average kinetic energy of the individual particles of a substance. (p. 46)

transition elements: elements in groups 3–12 in the periodic table, all of which are metals. (p. 99)

temperatura: medida de la energía cinética prome dio de las partículas individuales de una sustancia. (p. 46)

elementos de transición: elementos en los grupos 3-12 en la tabla periódica, todos los cuales son metales. (p. 99)

vaporization: the process by which a liquid changes into a gas. (pp. 50, 79)

viscosity: a liquid's resistance to flow. (p. 43)

vaporización: proceso mediante el cual un líquido se convierte en gas. (pp. 50, 79)

viscosidad: resistencia de un líquido al flujo. (p. 43)

Glossary/Glosario

Italic numbers = illustration/photo   **Bold numbers = vocabulary term**
lab = a page on which the entry is used in a lab
act = a page on which the entry is used in an activity

A

Actinides, 114, *115*
Activities, Applying Math, 59, 84; Applying Science, 27, 49, 103; Integrate Astronomy, 83; Integrate Earth Science, 29, 42; Integrate Health, 116; Integrate Life Science, 28, 81, 109; Integrate Physics, 16, 46, 114; Science Online, 19, 28, 43, 49, 51, 61, 81, 102, 116; Standardized Test Practice, 36–37, 68–69, 94–95, 124–125
Alkali metals, 105, *105*
Alkaline earth metals, 106, *106*
Alpha particles, 12, 13, *13*, 14, 16
Aluminum, *106*
Amalgam, 116
Americium, 114
Ammonia, 75, 108, *108*
Amorphous solids, 42, 47, *47*
Applying Math, Calculating Density, 59; Chapter Review, 35, 67, 93, 123; Converting Temperatures, 84; Section Review, 23, 29, 76, 87, 104
Applying Science, How can ice save oranges?, 49; What does *periodic* mean in the periodic table?, 103; What's the best way to desalt ocean water?, 27
Applying Skills, 44, 111, 116
Archimedes' principle, 59, *59*, 62–63 *lab*
Area, and pressure, 55, *55*
Argon, 111, *111*
Arsenic, 109
Astatine, 110
Atmospheric pressure, 55, 55–57, *56*
Atom(s), 8–17; history of, 8–17; mass number of, 21; model of, *9*, 9–17, *12, 14, 15, 15 lab;* nucleus of, *14*, 14–16, *15, 16*

Atomic mass, 22, *22*
Atomic number, 21

B

Balanced pressure, 56, *56*
Batteries, lithium, 105
Behavior, as physical property, 75, *75*
Beryllium, *106*
Blood, as mixture, 27, *27*, 28
Blood pressure, 61 *act*
Bohr, Niels, 16, 17
Boiling point, 50, *50*; as physical property, 74, 75
Boron, 106
Boron family, 106, *106*
Bromine, 75, 110
Buoyant force, 58, 58–59, *59*, 62–63 *lab*
Burning, 84, *84*

C

Calcium carbonate, 75
Californium-252, 114
Carbon, 107, *107*
Carbon dioxide, 25
Carbon group, 107, *107*
Carbon monoxide, 25
Catalysts, 113, *113*
Cathode-ray tube (CRT), 10, *10*, 11, *11*
Cerium, 114, *114*
Chemical changes, *80*, 80–85; color, 81, *81*; comparing to physical changes, 81 *lab*, 85, *85*; and energy, 82, *82*; recognizing, 81 *act*; reversing, 84, *84*; signs of, *81*, 81–84, *82, 83, 84*
Chemical formulas, 26
Chemical properties, 76, *76*, 77 *lab*
Chemistry, 8
Chlorine, 22, *22*, 110, *110*

Chlorophyll, 81
Chromium, 113
Classification, of elements, 19, *20*, 22–23
Cobalt, 112
Communicating Your Data, 24, 31, 53, 63, 77, 89, 119
Compound(s), *25*, **25**–26, *26*; comparing, 26 *lab*; formulas for, 26, *26*
Computers, and semiconductors, 107, *107*
Condensation, 48, **51,** *51*, 51 *act*, **79,** *79*
Conservation, of mass, 87, *87*
Crookes, William, 9, 10
Crystal, 41, *41*
Crystalline solids, 41, *41*, 47
Cycles, water, 53 *lab*

D

Dalton, John, 9, 20
Data Source, 118
Density, 59, 59 *act*
Dentistry, elements used in, 116
Deposition, 79
Desalination, 27 *act*
Design Your Own, Battle of the Toothpastes, 88–89; Design Your Own Ship, 62–63
Diamond, 107
Dissolving, as physical change, 79, *79*
Dry ice, 52, *52*
Ductility, 22

E

Electron(s), 11, 16–17, *17*
Electron cloud, 17, *17*
Element(s), 9, 18–23, 102 *act;* atomic mass of, 22, *22;* atomic number of, 21;

boron family of, 106, *106;* carbon group of, 107, *107;* classification of, 19, *20,* 22–23; halogens, 110, *110;* identifying characteristics of, 21–22; isotopes of, 21, *21;* metalloids, 23, 102, *102,* 106, 107, *107,* 109, 110; metals, 22, *22,* 102, *102, 105,* 105–106, *106,* 107, *107,* 112–116; new, 19 *act;* nitrogen group of, 108, *108;* noble gases, 110, 110–111, *111;* nonmetals, 23, *23,* 102, *102,* 107, *107,* 108, *108,* 109, 110, *110;* oxygen family of, 109, *109;* periodic table of, 19, *20, 21,* 24 *lab. See* Periodic table; radioactive, 114, *115;* representative, **99;** symbols for, 19, *20,* 104; synthetic, 18, **114,** *115;* transition, **99,** *112,* 112–116, *113, 114, 115*
Element keys, 103, *103*
Energy, and chemical changes, 82, *82;* thermal, *45,* **45**–46; types of, 46
Eruptions, volcanic, 70, *70*–71
Europium oxide, 114
Evaporation, *48, 50,* 50 *lab,* 50–51

Fertilizer, 108, *108*
Firefighting, foam for, 109, *109*
Fireworks, 80, *80*
Flint, 114, *114*
Fluids, 54–61. *See* Liquid(s). *See also* Gas(es); and Archimedes' principle, 59, *59,* 62–63 *lab;* and buoyant force, 58, 58–59, *59,* 62–63 *lab;* and density, 59, 59 *act;* and Pascal's principle, *60,* 60–61, *61;* and pressure, 54–58
Fluoride, 116
Fluorine, 110
Foam, for firefighting, 109, *109*
Foldables, 7, 39, 71, 97
Force(s), 54; and area, 55, *55;* buoyant, 58, **58**–59, *59,* 62–63 *lab;* measurement of, 54; and pressure, 54–58
Force pumps, 61, *61*
Formulas, chemical, 26; for compounds, 26, *26*

Freezing, 39 *lab, 48,* **49,** 79, *79*
Freezing point, 49, 49 *act,* 75
Fusion, 115

Gallium, 106
Gas(es), 44, *44. See also* Fluids; and chemical changes, 83, *83;* condensation of, *48,* 51, *51,* 51 *act;* pressure of, *57,* 57–58, *58*
Germanium, 107
Glass, 47, *47,* 107
Graphite, 107
Group, 19, **99**

Halogens, 110, *110*
Health, and heavy metals, 118–119 *lab;* and mercury, 116 *act*
Heart, 61, *61*
Heat, 46; specific, 47, *47;* and temperature, 46–47
Heavy metals, 107, *107,* 113, 118–119 *lab*
Helium, 110, 111, *111*
Hemoglobin, 112
Heterogeneous mixtures, 29
Homogeneous mixtures, 29
Hydraulic systems, 60, *60*
Hydrogen peroxide, 26, *26*
Hydrogen v. helium, 110

Ice, dry, 52, *52*
Inner transition elements, 114, *114, 115*
Integrate Astronomy, meteoroid, 83
Integrate Earth Science, freshwater, 42; rocks and minerals, 29
Integrate Health, dentistry and dental materials, 116
Integrate Life Science, blood as a mixture, 28; poison buildup, 109; signs of chemical changes, 81
Integrate Physics, bright lights, 114; quantum theory, 16; types of energy, 46

International Union of Pure and Applied Chemistry (IUPAC), 104
Iodine, 75, 110
Iridium, 113
Iron, 112, 114
Iron triad, 112, *112*
Isotopes, 21, *21*

Journal, 6, 38, 70, 96

Kilopascal (kPa), 54
Krypton, 111

Lab(s), Design Your Own, 62–63, 88–89; Elements and the Periodic Table, 24; Finding the Difference, 77; Launch Labs, 7, 39, 71, 97; Mini Labs, 26, 50, 74, 75, 99; Mystery Mixture, 30–31; Try at Home Mini Labs, 15, 57, 81; Use the Internet, 118–119; Water Cycle, 53
Lanthanides, 114, *114*
Lanthanum, 114, *114*
Launch Labs, Changing Face of a Volcano, 71; Experiment with a Freezing Liquid, 39; Make a Model of a Periodic Pattern, 97; Model the Unseen, 7
Lava, 71 *lab*
Lavoisier, Antoine, 20, 87
Lawrencium, 114
Law(s), of conservation of mass, **87,** *87*
Lead, 107, *107,* 114
Leaves, changing colors of, 81, *81*
Lightbulb, 113, *113*
Liquid(s), 42, 42–43, *43. See also* Fluids; freezing, 39 *lab, 48,* 49; and surface tension, 43, *43;* vaporization of, *48,* 50 *lab, 50,* 50–51; viscosity of, 43
Lithium, 105
Lodestone, 75, *75*
Luster, 22
Lutetium, 114

Index

M

Magnesium, *106*
Magnetic properties, 75, *75*, 112
Malleability, 20, *20*
Mass, conservation of, 87, *87*
Mass number, 21
Materials, semiconductors, 107, *107*
Matter, 40. *See also* States of matter; ancient views of, 8, 32; appearance of, 73, *73*; compounds, 25, *25*–26, *26*; describing, 72–77, 77 *lab*; elements in, 18–23, 24 *lab*
Measurement, of force, 54; of properties, 74 *lab*; of weight, *74*
Meitner, Lise, 104
Melting, 47, *47, 48,* 79, *79*
Melting point, 47, 74, 75
Mendeleev, Dmitri, 20, 98, *98,* 99
Mercury, 113, 116, 116 *act*
Metal(s), 22, *22,* **102;** alkali, **105,** *105;* alkaline earth, **106,** *106;* as catalysts, 113, *113;* heavy, 107, *107,* 113, 118–119 *lab;* iron triad, 112, *112;* misch, 114, *114;* on periodic table, 102, *102, 105,* 105–106, *106,* 107, *107,* 112–116; transition, *112,* 112–116, *113, 114, 115*
Metalloids, 23, 102, *102,* 106, 107, *107,* 109, 110
Meteoroid, 83
Mineral(s), 29
Mini Labs, Comparing Compounds, 26; Designing a Periodic Table, 99; Identifying an Unknown Substance, 75; Measuring Properties, 74; Observing Vaporization, 50
Misch metal, 114, *114*
Mixtures, 27–31, 28 *act;* blood as, 27, *27,* 28; heterogeneous, 29; homogeneous, 29; identifying, 30–31 *lab;* separating, 28, *28*
Model(s), of atom, *9,* 9–17, *12, 14, 15,* 15 *lab;* of unseen, 7 *lab*
Moseley, Henry, 99

N

National Geographic Visualizing, The Periodic Table, 20; Recycling, 86; States of Matter, *48;* Synthetic Elements, *115*
Neodymium, 114, *114*
Neon, *73,* 111, *111*
Neutron(s), 15
Newton (unit of force), 54
Nickel, 112
Nitrogen, 108, *108*
Nitrogen group, 108, *108*
Noble gases, *110,* **110**–111, *111*
Nonmetals, 23, *23,* **102,** *102,* 107, *107,* 108, *108,* 109, 110, *110*
Nucleus, *14,* 14–16, *15, 16*

O

Ocean water, desalination of, 27 *act;* salt in, 27 *act*
Odor, and chemical changes, 83
Oil (petroleum), *73*
Oops! Accidents in Science, Incredible Stretching Goo, 64
Osmium, 113
Oxygen, on periodic table, 109, *109*
Oxygen family, 109, *109*
Ozone, 109

P

Palladium, 113
Particle(s), alpha, 12–13, *13,* 14, 16; charged, 11–13
Particle accelerator, 115, *115*
Pascal (Pa), 54
Pascal's principle, 60, *60*–61, *61*
Period, 19, 99
Periodic pattern, making models of, 97 *lab*
Periodic table, 19, *20, 21,* 24 *lab,* 96–117, *100–101;* boron family on, 106, *106;* carbon group on, 107, *107;* designing, 99 *lab;* development of, *98,* 98–99; element keys on, 103, *103;* halogens on, 110, *110;* metalloids on, 102, *102,* 106, 107, *107,* 109, 110; metals on, 102, *102, 105,* 105–106, *106, 107,* 112–116; nitrogen group on, 108, *108;* noble gases on, *110,* 110–111, *111;* nonmetals on, 102, *102,* 107, *107,* 108, *108,* 109, 110, *110;* oxygen family on, 109, *109;* symbols for elements on, 104; zones on, 99, *99, 102, 102*–104, *103*
Phosphorus, 108, *108*
Physical changes, 78, **78**–79, *79;* comparing to chemical changes, 81 *lab,* 85, *85;* reversing, 84
Physical properties, *72,* **72**–75, *74,* 77 *lab;* appearance, 73, *73;* behavior, 75, *75;* boiling point, 74, 75; magnetic, 75, *75,* 112; measuring, 74 *lab;* melting point, 74, 75; state, 73, *73*
Pigment, 81
Pistons, 60, *60*
Plant(s), chlorophyll in, 81; leaves of, 81, *81*
Plasma, 43 *act*
Platinum, 113
Platinum group, 113
Plutonium, 114
Poisons, 99
Polonium, 109
Potassium, 105
Potassium hydroxide, 75
Pressure, 54, 54–58; and area, 55, *55;* atmospheric, 55, 55–57, *56;* balanced, 56, *56;* and force, 54–58; of gas, *57,* 57–58, *58;* and temperature, 58, *58;* and volume, 57, *57*
Properties, chemical, **76,** *76,* 77 *lab;* comparing, 71 *lab;* magnetic, 75, *75,* 112; physical. *See* Physical properties
Protactinium, 114
Proton(s), 14

Q

Quantum theory, 16
Quartz, 107

R

Radioactive elements, 114, *115*

Radon, 111
Reading Check, 10, 12, 14, 15, 21, 26, 27, 40, 41, 42, 44, 46, 51, 55, 58, 72, 75, 80, 84, 102, 106, 108, 110, 111, 112, 114, 116
Reading Strategies, 8A, 40A, 72A, 98A
Real-World Questions, 24, 30, 53, 62, 77, 88, 118
Recycling, *86,* 90, *90*
Representative elements, 99
Rhodium, 113
Rock(s), 29
Rust, 80, *80*
Ruthenium, 113
Rutherford, Ernest, 12–13, 14

Salt(s), 105, 110, *110;* and chemical changes, 82, *82;* crystal structure of, 41, *41;* physical properties of, 75
Sand, 107
Science and History, Ancient Views of Matter, 32
Science and Language Arts, "Anansi Tries to Steal All the Wisdom in the World," 120
Science Online, blood pressure, 61; condensation, 51; elements, 102; freezing point study, 49; health risks, 116; mixtures, 28; new elements, 19; plasma, 43; recognizing chemical changes, 81
Science Stats, Strange Changes, 90
Scientific Methods, 24, 30–31, 53, 62–63, 77, 88–89, 117, 118–119; Analyze Your Data, 31, 63, 89, 119; Conclude and Apply, 24, 31, 53, 63, 77, 89, 119; Follow Your Plan, 63, 119; Form a Hypothesis, 62, 88; Make a Plan, 63, 119; Test Your Hypothesis, 63, 89
Selenium, 109, *109*
Semiconductors, 107, *107*
Shape, changes of, 78, *78*
Ship, designing, 62–63
Silicon, 107, *107*
Silver tarnish, 80, *80*
Smell, and chemical changes, 83

Sodium chloride, 41, *41,* 75, 105, 110, *110. See also* Salt(s)
Solid(s), *41,* **41**–42; amorphous, 42, 47, *47;* and chemical changes, 83, *83;* crystalline, 41, *41,* 47; melting, 47, *47, 48;* sublimation of, 52, *52*
Space shuttle, *25*
Specific heat, 47, *47*
Spring scale, *74*
Standardized Test Practice, 36–37, 68–69, 94–95, 124–125
States of matter, 38–63, *40;* changes of, 45–53, 53 *lab;* and condensation, *48,* 51, *51,* 51 *lab;* and evaporation, *48, 50,* 50 *lab,* 50–51; fluids, 54–61, *58, 59,* 59 *act, 60, 61,* 62–63 *lab;* and freezing, 39 *lab, 48,* 49; gases, 44, *44;* liquids, *42,* 42–43, *43;* and melting, 47, *47, 48;* as physical change, 79, *79;* as physical property, 73, *73;* and pressure, 54–58; solids, *41,* 41–42; and sublimation, 52, *52;* and vaporization, *48, 50,* 50 *lab,* 50–51
Steel, 112, *112*
Study Guide, 33, 65, 91, 121
Sublimation, 52, *52,* **79**
Substance, 25
Sugars, dissolving, 79, *79*
Sulfur, 109
Sulfuric acid, 109
Surface tension, 43, *43*
Symbols, for elements, 19, *20,* 104
Synthetic elements, 18, **114,** *115*

Tarnish, 80, *80*
Technology, cathode-ray tube (CRT), 10, *10,* 11, *11;* computers, 107, *107;* in dentistry, 116; fireworks, 80, *80;* lightbulb, 113, *113;* particle accelerator, 115, *115;* semiconductors, 107, *107;* space shuttle, *25;* spring scale, *74;* synthetic elements, 114, *115;* Tevatron, *18*
Teeth, 88–89 *lab. See also* Dentistry
Tellurium, 109
Temperature, 46, *46;* and heat, 46–47; and pressure, 58, *58*

Tevatron, *18*
Thermal energy, 45, **45**–46
Thomson, J. J., 11–12, 14
Thorium, 114
TIME, Science and History, 32
Tin, 107
Toothpastes, comparing, 88–89 *lab*
Transition elements, 99, *112,* 112–116, *113;* in dentistry, 116; inner, 114, *114, 115*
Try at Home MiniLabs, Comparing Changes, 81; Modeling the Nuclear Atom, 15; Predicting a Waterfall, 57
Tungsten, 113, *113*

Unknown, finding, 59 *act,* 84 *act*
Uranium, 114, 115
Use the Internet, Health Risks from Heavy Metals, 118–119

Vapor, 44
Vaporization, *48, 50,* 50 *lab,* **50**–51, *79, 79*
Viscosity, 43
Volcanoes, changing face of, 71 *lab;* eruptions of, 70, *70–71*
Volume, and pressure, 57, *57*

Water, boiling point of, 50, *50;* changes of state of, 79, *79;* freshwater, 42; melting point of, 47; physical properties of, 75
Water cycle, 53 *lab*
Waterfalls, 57 *lab*
Wave(s), electron as, 17
Weight, measuring, *74*

Xenon, 111

Yttrium oxide, 114

Index

Magnification Key: Magnifications listed are the magnifications at which images were originally photographed.
LM–Light Microscope
SEM–Scanning Electron Microscope
TEM–Transmission Electron Microscope

Acknowledgments: Glencoe would like to acknowledge the artists and agencies who participated in illustrating this program: Absolute Science Illustration; Andrew Evansen; Argosy; Articulate Graphics; Craig Attebery represented by Frank & Jeff Lavaty; CHK America; John Edwards and Associates; Gagliano Graphics; Pedro Julio Gonzalez represented by Melissa Turk & The Artist Network; Robert Hynes represented by Mendola Ltd.; Morgan Cain & Associates; JTH Illustration; Laurie O'Keefe; Matthew Pippin represented by Beranbaum Artist's Representative; Precision Graphics; Publisher's Art; Rolin Graphics, Inc.; Wendy Smith represented by Melissa Turk & The Artist Network; Kevin Torline represented by Berendsen and Associates, Inc.; WILDlife ART; Phil Wilson represented by Cliff Knecht Artist Representative; Zoo Botanica.

Photo Credits

Cover Roine Magnusson/Stone; **i ii** Roine Magnusson/Stone; **iv** (bkgd)John Evans, (inset)Roine Magnusson/Stone; **v** (t)PhotoDisc, (b)John Evans; **vi** (l)John Evans, (r)Geoff Butler; **vii** (l)John Evans, (r)PhotoDisc; **viii** PhotoDisc; **ix** Aaron Haupt Photography; **x** Roger Ressmeyer/CORBIS; **xi** SuperStock; **xii** Richard Megna/Fundamental Photographs; **1** Joseph Sohm/ChromoSohm, Inc./CORBIS; **2** Charles O'Rear/CORBIS; **3** (t)Charlie Varley/Sipa, (b)Philip Hayson/Photo Researchers; **4** Charles O'Rear/CORBIS; **5** Charlie Varley/SIPA; **6–7** courtesy IBM; **8** EyeWire; **9** (tl)Culver Pictures/PictureQuest, (tr)E.A. Heiniger/Photo Researchers, (b)Andy Roberts/Stone/Getty Images; **10** Elena Rooraid/PhotoEdit, Inc.; **11** (t)L.S. Stepanowicz/Panographics, (b)Skip Comer; **12** Aaron Haupt; **16** Fraser Hall/Robert Harding Picture Library; **18** Fermi National Accelerator Laboratory/Science Photo Library/Photo Researchers; **19** Tom Stewart/The Stock Market/CORBIS; **20** (bkgd tr bl)Bettmann/CORBIS, (br)New York Public Library, General Research Division, Astor, Lenox, and Tilden Foundations; **22** Emmanuel Scorcelletti/Liaison Agency/Getty Images; **24** Doug Martin; **25** NASA; **26** Mark Burnett; **27** Klaus Guldbrandsen/Science Photo Library/Photo Researchers; **28** (tl)Mark Thayer, (tr)CORBIS, (bl)Kenneth Mengay/Liaison Agency/Getty Images, (bc)Arthur Hill/Visuals Unlimited, (br)RMIP/Richard Haynes; **28–29** KS Studios; **30** (t)Mark Burnett, (b)Michael Newman/PhotoEdit, Inc.; **32** (tl)Robert Essel/The Stock Market/CORBIS, (tr)John Eastcott & Yva Momatiuk/DRK Photo, (cr)Diaphor Agency/Index Stock, (bl)Ame Hodalic/CORBIS, (br)TIME; **38–39** Roger Ressmeyer/CORBIS; **40** Layne Kennedy/CORBIS; **41** (t)Telegraph Colour Library/FPG/Getty Images, (b)Paul Silverman/Fundamental Photographs; **42** Bill Aron/PhotoEdit, Inc.; **43** (l)John Serrao/Photo Researchers, (r)H. Richard Johnston; **44** Tom Tracy/Photo Network/PictureQuest; **45** Annie Griffiths Belt/CORBIS; **46** Amanita Pictures; **47** (t)David Weintraub/Stock Boston, (b)James L. Amos/Peter Arnold, Inc.; **48** Dave King/DK Images; **49** Joseph Sohm/ChromoSohm, Inc./CORBIS; **50** Michael Dalton/Fundamental Photographs; **51** Swarthout & Associates/The Stock Market/CORBIS; **52** Tony Freeman/PhotoEdit, Inc. **54** David Young-Wolff/PhotoEdit, Inc.; **55** (t)Joshua Ets-Hokin/PhotoDisc, (b)Richard Hutchings; **56** Robbie Jack/CORBIS; **58** A. Ramey/Stock Boston; **59** Mark Burnett; **60** (t)Tony Freeman/PhotoEdit, Inc., (b)Stephen Simpson/FPG/Getty Images; **62** (t)Lester Lefkowitz/The Stock Market/CORBIS, (b)Bob Daemmrich; **63** Bob Daemmrich; **64** Daniel Belknap; **65** (l)Andrew Ward/Life File/PhotoDisc, (r)NASA/TRACE; **67** Mark Burnett; **69** Joshua Ets-Hokin/PhotoDisc; **70–71** James L. Amos/CORBIS; **72** Fred Habegger from Grant Heilman; **73** (tr)David Nunuk/Science Photo Library/Photo Researchers, (cr)Mark Burnett, (bl)David Schultz/Stone/Getty Images, (bc)SuperStock, (br)Kent Knudson/PhotoDisc; **74** KS Studios; **75** Gary Retherford/Photo Researchers; **76** (l)Peter Steiner/The Stock Market/CORBIS, (c)Tom & DeeAnn McCarthy/The Stock Market/CORBIS, (r)SuperStock; **77** Timothy Fuller; **78** (l)Gay Bumgarner/Stone/Getty Images, (r)A. Goldsmith/The Stock Market/CORBIS; **79** (t)Matt Meadows, (others)Richard Megna/Fundamental Photographs; **80** (tl)Ed Pritchard/Stone/Getty Images, (cl bl)Kip Peticolas/Fundamental Photographs, (tr br)Richard Megna/Fundamental Photographs; **81** Rich Iwasaki/Stone/Getty Images; **82** (t)Matt Meadows, (bc)Layne Kennedy/CORBIS, (bl br)Runk/Schoenberger from Grant Heilman; **83** (t)Amanita Pictures, (b)Richard Megna/Fundamental Photographs; **84** Anthony Cooper/Ecoscene/CORBIS; **85** (tl)Russell Illig/PhotoDisc, (tcl)John D. Cunningham/Visuals Unlimited, (tcr)Coco McCoy/Rainbow/PictureQuest, (bl)Bonnie Kamin/PhotoEdit, Inc., (tr br)SuperStock; **86** (t)Grantpix/Photo Researchers, (c)Mark Sherman/Photo Network/PictureQuest, (bl)Sculpture by Maya Lin, courtesy Wexner Center for the Arts, Ohio State Univ., photo by Darnell Lautt, (br)Rainbow/PictureQuest; **87** Mark Burnett; **88 89** Matt Meadows; **90** (l)Susan Kinast/Foodpix/Getty Images, (r)Michael Newman/PhotoEdit, Inc.; **91** (l)C. Squared Studios/PhotoDisc, (r)Kip Peticolas/Fundamental Photographs; **94** Elaine Shay; **95** (l)Kurt Scholz/SuperStock, (r)CORBIS; **96–97** Jim Corwin/Index Stock; **98** Stamp from the collection of Prof. C.M. Lang, photo by Gary Shulfer, University of WI Stevens Point; **102** (tl)Tom Pantages, (tr)Elaine Shay, (bl)Paul Silverman/Fundamental Photographs; **105** Amanita Pictures; **106** (l)Joail Hans Stern/Liaison Agency/Getty Images, (r)Leonard Freed/Magnum/PictureQuest; **107** (l)David Young-Wolff/PhotoEdit/PictureQuest, (c)Jane Sapinsky/The Stock Market/CORBIS, (r)Dan McCoy/Rainbow/PictureQuest; **108** (t)George Hall/CORBIS, (b)Aaron Haupt; **109** SuperStock; **110** (t)Don Farrall/PhotoDisc, (b)Matt Meadows; **111** (l)file photo, (r)Bill Freund/CORBIS; **112** CORBIS; **113** (t)Geoff Butler, (b)Royalty-Free/CORBIS; **114** Amanita Pictures; **115** (l)Achim Zschau, (r)Ted Streshinsky/CORBIS; **118** Robert Essel NYC/CORBIS; **119** Mark Burnett; **120** Tim Flach/Stone/Getty Images; **121** (l)Yoav Levy/PhotoTake NYC/PictureQuest, (r)Louvre, Paris/Bridgeman Art Library, London/New York; **122** Matt Meadows; **126** PhotoDisc; **128** Tom Pantages; **132** Michell D. Bridwell/PhotoEdit, Inc.; **133** (t)Mark Burnett, (b)Dominic Oldershaw; **134** StudiOhio; **135** Timothy Fuller; **136** Aaron Haupt; **138** KS Studios; **139** Matt Meadows; **140** Mark Burnett; **142** Amanita Pictures; **143** Bob Daemmrich; **145** Davis Barber/PhotoEdit, Inc.

PERIODIC TABLE OF THE ELEMENTS

Columns of elements are called groups. Elements in the same group have similar chemical properties.

Gas
Liquid
Solid
Synthetic

Element — Hydrogen
Atomic number — 1
Symbol — H
Atomic mass — 1.008

State of matter

The first three symbols tell you the state of matter of the element at room temperature. The fourth symbol identifies elements that are not present in significant amounts on Earth. Useful amounts are made synthetically.

1	2	3	4	5	6	7	8	9
1 Hydrogen 1 **H** 1.008								
2 Lithium 3 **Li** 6.941	Beryllium 4 **Be** 9.012							
3 Sodium 11 **Na** 22.990	Magnesium 12 **Mg** 24.305							
4 Potassium 19 **K** 39.098	Calcium 20 **Ca** 40.078	Scandium 21 **Sc** 44.956	Titanium 22 **Ti** 47.867	Vanadium 23 **V** 50.942	Chromium 24 **Cr** 51.996	Manganese 25 **Mn** 54.938	Iron 26 **Fe** 55.845	Cobalt 27 **Co** 58.933
5 Rubidium 37 **Rb** 85.468	Strontium 38 **Sr** 87.62	Yttrium 39 **Y** 88.906	Zirconium 40 **Zr** 91.224	Niobium 41 **Nb** 92.906	Molybdenum 42 **Mo** 95.94	Technetium 43 **Tc** (98)	Ruthenium 44 **Ru** 101.07	Rhodium 45 **Rh** 102.906
6 Cesium 55 **Cs** 132.905	Barium 56 **Ba** 137.327	Lanthanum 57 **La** 138.906	Hafnium 72 **Hf** 178.49	Tantalum 73 **Ta** 180.948	Tungsten 74 **W** 183.84	Rhenium 75 **Re** 186.207	Osmium 76 **Os** 190.23	Iridium 77 **Ir** 192.217
7 Francium 87 **Fr** (223)	Radium 88 **Ra** (226)	Actinium 89 **Ac** (227)	Rutherfordium 104 **Rf** (261)	Dubnium 105 **Db** (262)	Seaborgium 106 **Sg** (266)	Bohrium 107 **Bh** (264)	Hassium 108 **Hs** (277)	Meitnerium 109 **Mt** (268)

The number in parentheses is the mass number of the longest-lived isotope for that element.

Rows of elements are called periods. Atomic number increases across a period.

The arrow shows where these elements would fit into the periodic table. They are moved to the bottom of the table to save space.

Lanthanide series

Cerium 58 **Ce** 140.116	Praseodymium 59 **Pr** 140.908	Neodymium 60 **Nd** 144.24	Promethium 61 **Pm** (145)	Samarium 62 **Sm** 150.36

Actinide series

Thorium 90 **Th** 232.038	Protactinium 91 **Pa** 231.036	Uranium 92 **U** 238.029	Neptunium 93 **Np** (237)	Plutonium 94 **Pu** (244)